Phenotypic and Genotypic Diversity of Rhizobia

Authored by

Neelawan Pongsilp

Department of Microbiology
Faculty of Science
Silpakorn University
Sanam-Chandra Palace Campus
Nakhon Pathom
Thailand

DEDICATION

This book is dedicated to my mother, Yingpan Pongsilp, as well as my supervisors, Prof. Nantakorn Boonkerd (School of Biotechnology, Suranaree University of Technology, Thailand) and Prof. Michael J. Sadowsky (Department of Soil, Water and Climate, University of Minnesota, USA). A special thank goes to Dr. Pongrawee Nimnoi, all research colleagues and friends.

Neelawan Pongsilp, Ph.D.

Department of Microbiology
Faculty of Science
Silpakorn University
Sanam-Chandra Palace Campus
Nakhon Pathom
Thailand

CONTENTS

Foreword   *i*

Preface   *iii*

CHAPTERS

1. **Classification and Taxonomy of Rhizobia**   **3**

2. **Phenotypic Diversity of Rhizobia Assessed by Numerical Analysis, Enzyme Pattern and Serological Study**   **49**

3. **Symbiotic Variation and Plant-Growth-Promoting Traits of Rhizobia**   **73**

4. **Genotypic Diversity of Rhizobia Assessed by Plasmid Profile**   **93**

5. **Genotypic Diversity of Rhizobia Assessed by Polymerase Chain Reaction (PCR) Fingerprinting**   **107**

6. **Genotypic Diversity of Rhizobia Assessed by Polymerase Chain Reaction-Restriction Fragment Length Polymorphism (PCR-RFLP)**   **125**

7. **Genotypic Diversity of Rhizobia Assessed by Restriction Fragment Length Polymorphism (RFLP)**   **136**

8. **Genotypic Diversity of Rhizobia Assessed by Sequence Analysis**   **154**

9. **Application of Rhizobia in Agriculture**   **176**

    Index   **188**

FOREWORD

It is such an honor and I have a great pleasure to write a foreword for this special book, "PHENOTYPIC AND GENOTYPIC DIVERSITY OF RHIZOBIA", written by Dr. Neelawan Pongsilp.

Rhizobia are defined as bacteria that induce nitrogen-fixing nodules on roots and stems of legumes. Members of rhizobia have expanded from the only one genus, *Rhizobium*, in the early 80's to 17 genera identified up to date. The classification and taxonomy of rhizobia have been revised periodically. Many rhizobial strains have been moved into the other genera and species. Novel genera and species of rhizobia in the alpha (α), beta (β) and gamma (γ) subclasses of Proteobacteria have been proposed as well. The diversity of rhizobia has been assessed based on phenotypic and genotypic characteristics. The genetic relationships among rhizobial populations are of interest because they can provide the information on the gene transfer and the adaptation of bacteria to environments. Despite increasing studies on rhizobial diversity and their importance in sustainable agriculture world-wide, the gathering information is quite rare.

This book provides valuable information. It covers the main concepts in classification and taxonomy of rhizobia. It brings the categories and techniques used to examine the phenotypic and genotypic diversity of rhizobia, from principle to application for rhizobial members. The techniques described in the book include numerical analysis, enzyme pattern, serological study, plasmid profile, PCR fingerprinting, PCR-RFLP, RFLP, IS fingerprint and sequence analysis. The discriminating power and limitations of these techniques are discussed and compared. The plant-growth promoting abilities of rhizobia are also provided. In this book, the information is gathered to well-organized and easily comparable formats. It rises up the new aspects on the symbiotic promiscuity and genetic relatedness of rhizobia.

The author, Dr. Neelawan Pongsilp, has a vast experience in bacterial diversity and community, especially rhizobia. Her research articles published in the peer-reviewed journals prove her well qualified for writing on this topic. She pushes the Rhizobia world forward with her slender frame. For years, I've admired her devotion to researches and got inspired by her hard work, professional skills and opinions. As I know that she's made an excellent effort in writing this book, she achieves a good progress in providing a useful resource of knowledge. This book will benefit readers in microbiological and molecular biological fields.

Pongrawee Nimnoi, Ph.D.
Department of Microbiology
Faculty of Liberal Arts and Sciences
Kasetsart University
Nakhon Pathom
Thailand

PREFACE

I would like to take this opportunity to mention briefly and express my aim about this book.

Rhizobia are composed of specific groups of bacteria that have the ability to induce symbiotic nitrogen-fixing nodules on the roots or stems of leguminous plants. Rhizobia have attracted a great attention for more than 4 decades because of their enormous agricultural and economic values in sustainable agriculture. Up to present time, rhizobia are classified in to diverse taxonomic groups, distributing in 17 genera (118 species) of the alpha (α), beta (β) and gamma (γ) subclasses of Proteobacteria. The classification of rhizobia is becoming increasingly complex and is revised periodically because of the discovery of new rhizobial members in the other genera and species, the proposal of novel rhizobial genera and species as well as the criteria used for classification. An assessment of rhizobial diversity provides pivotal information in understanding the horizontal gene transfer among bacterial genera and species, the bacterial evolution and the symbiotic effectiveness.

The main concepts of this book present the background knowledge of rhizobia, the broad perspective on rhizobial diversity, the information on characteristics specific to each group of rhizobia, the relation between rhizobial groups and genetic factors contributed to rhizobial diversity, the contemporary methods for examination of rhizobial diversity, the plant-growth-promoting traits of rhizobia and the application of rhizobia. In the first chapter, classification and taxonomy of rhizobia are provided with general information of rhizobia, criteria used in classification of rhizobia, symbiotic gene transfer among rhizobial members, non-symbiotic rhizobia and symbiotic promiscuity of rhizobia. A compiled list of rhizobial species with their hosts is also included. The second chapter describes phenotypic diversity of rhizobia based on numerical analysis, enzyme production and serogroups. The third chapter is about symbiotic variation and plant-growth promoting traits of rhizobia including production of phytohormones, siderophores and ammonia, phosphate-solubilizing activity and production of ACC deaminase. The forth to eight chapters contribute to genotypic diversity of rhizobia, separating molecular techniques used for examination. The techniques listed in this book are the most frequently used ones including plasmid profile, PCR fingerprinting, PCR-RFLP, RFLP and sequence analysis. The details of each technique include principle concept, application to rhizobia, advantages and limitations as well as comparison of resolution level and consistency. The ninth chapter focuses on the application of rhizobia. The uses of rhizobia in agriculture and bioremediation, co-inoculations of rhizobia and other plant-growth-promoting rhizobacteria as well as influence of inoculants on genotypic diversity of indigenous rhizobia are involved.

I hope the book can bring the interesting information and ideas to readers world-wide. From background to specific sections, it may facilitate the proper and efficient experiments related to rhizobial and bacterial diversity.

ACKNOWLEDGEMENT

Declared none.

CONFLICT OF INTEREST

The author confirms that this chapter content has no conflicts of interest.

Neelawan Pongsilp, Ph.D.
Department of Microbiology
Faculty of Science
Silpakorn University-Sanam Chandra Palace Campus
Nakhon Pathom
Thailand

CHAPTER 1

Classification and Taxonomy of Rhizobia

Neelawan Pongsilp[*]

Department of Microbiology, Faculty of Science, Silpakorn University, Nakhon Pathom, Thailand

Abstract: Rhizobia are composed of specific groups of bacteria that have the ability to induce symbiotic nitrogen-fixing nodules on the roots or stems of leguminous plants. Rhizobia are of enormous agricultural and economic values because they provide the major source of nitrogen input in agricultural soils. The classification of rhizobia is becoming increasingly complex and is revised periodically because of new findings that propose new genera and new species. Up to the present time, rhizobia are distributed in 17 genera including *Azorhizobium*, *Bradyrhizobium*, *Burkholderia*, *Cupriavidus*, *Devosia*, *Ensifer*, *Herbaspirillum*, *Mesorhizobium*, *Methylobacterium*, *Microvirga*, *Ochrobactrum*, *Phyllobacterium*, *Pseudomonas*, *Ralstonia*, *Rhizobium*, *Shinella* and *Sinorhizobium*. Rhizobial diversity has been revealed by several methodologies, providing valuable information about bacterial genotypes that are well adapted to a certain environment.

Keywords: Biological nitrogen fixation (BNF), Classification, Host specificity, Non-symbiotic rhizobia, Rhizobia, Rhizobial diversity, Symbiosis, Symbiotic gene, Symbiotic promiscuity, Taxonomy.

1.1. GENERAL INFORMATION OF RHIZOBIA

Rhizobia are composed of specific groups of bacteria that have the ability to induce symbiotic nitrogen-fixing nodules on the roots or stems of leguminous plants. In the presence of available nitrogen, they can exist as free-living soil saprophytes. At a particular condition (in the absence of available nitrogen), these bacteria interact with the roots or stems of leguminous plants, inducing the formation of nodules in which the fixation of atmospheric nitrogen occurs [1]. The intracellular bacteria, termed as "bacteroids", are surrounded by a plant plasmalemma-derived membrane and have a physiological state that is different from the free-living state. Bacteroids convert atmospheric nitrogen into ammonia,

*Address correspondence to Neelawan Pongsilp: Department of Microbiology, Faculty of Science, Silpakorn University, Nakhon Pathom, Thailand; Tel: +66-34-245337; Fax: +66-34-245336-37; E-mail: neelawan@su.ac.th

providing the nitrogen requirements of both rhizobia and their host plants. In return, rhizobia receive a carbon source, typically dicarboxylates and other nutrients from the plants [2, 3]. This biological nitrogen fixation (BNF) represents the major source of nitrogen input in agricultural soils. The major nitrogen-fixing systems are the symbiotic systems which can play a significant role in improving the fertility and productivity of low-nitrogen soils [4]. Consequently, rhizobia are of enormous agricultural and economic values [5]. The host plants of rhizobia, leguminous tree species, are classified in the family Fabaceae which is the third largest of angiosperms with approximately 650 genera and 20,000 species [6]. These plants are both abundant and diverse in tropical forests [7]. It was once believed that rhizobia-legume symbioses had a stringent host specificity. This means that only closely related legumes can be nodulated by a particular rhizobial strain. Up to the present time, many legumes have been found to be nodulated by several rhizobial species which belong to different taxonomic groups.

1.2. CRITERIA USED IN CLASSIFICATION OF RHIZOBIA

The genus *Rhizobium* was firstly proposed for bacteria that have the ability to induce root nodules of legumes. The second genus, *Bradyrhizobium*, was emerged as the genus of the slow-growing and alkaline-producing rhizobial strains. Both genera previously belonged to the family Rhizobiaceae and the classification was based on growth rate, production of acid or alkaline and host specificity [8, 9]. The classification of rhizobia is becoming increasingly complex and is revised periodically because of new findings that propose new genera and new species. DNA homology values, guanine-cytosine (GC) content, sequence homologies of multiple loci (such as small subunit ribosomal RNA gene (16S rDNA), house keeping genes and symbiotic genes), locations of symbiotic genes and phenotypic characteristics provide more and deeper information for the classification of rhizobia. Some species in genera *Rhizobium* and *Bradyrhizobium* were later moved into new genera based on phylogenetic analyses. *Sinorhizobium* was proposed as the new genus of the fast-growing soybean rhizobia. This genus is separated from *Rhizobium* and *Bradyrhizobium* based on GC content, DNA-DNA hybridization data, numerical taxonomy, serological analysis data, composition of extracellular gum, bacteriophage typing data and soluble protein pattern [10]. However, *Rhizobium* and *Sinorhizobium* are phylogenetically fairly closely related to each other [11]. Recently, the Judicial Commission of the International Committee on Systematics of Prokaryotes decided to transfer some species of the genus *Sinorhizobium* to the genus *Ensifer* [12]. The first species of *Azorhizobium*, *Azo. caulinodans*, was described on the basis of nodule formation on both stems and roots of a semi-aquatic tropical legume, *Sesbania rostrata* [13]. *Allorhizobium undicola* was proposed for a species which is capable of efficient nitrogen-fixing symbiosis with *Neptunia natans* [14]. *All. undicola* was later merged into the genus *Rhizobium* as *Rhi. undicola* [15], therefore the genus *Allorhizobium* is not currently valid. Based on phylogenetic relationships constructed from sequences of 16S rRNA gene, several members were removed from the genus *Rhizobium* to establish the new genus *Mesorhizobium* with the first 5 species including *Mes. ciceri*, *Mes. huakuii*, *Mes. loti*, *Mes. mediterraneum* and *Mes. tianshanense* [16]. The aquatic bacterium, *Blastobacter denitrificans*, was reported as a nitrogen-fixing symbiont of *Aeschynomene indica* (Indian jointvetch) [17]. This bacterium was later transferred to *Bradyrhizobium denitrificans* [18]. Although the genera *Rhizobium*, *Bradyrhizobium*, *Sinorhizobium*, *Azorhizobium* and *Mesorhizobium* were originally proposed as rhizobial genera, not all species have been verified for their nodule formation ability. Besides these genera, the nodule formation ability and the presence of symbiotic genes have been discovered among different genera of soil bacteria.

1.3. CLASSIFICATION AND TAXONOMY OF RHIZOBIA

Five genera (*Rhizobium*, *Bradyrhizobium*, *Sinorhizobium*, *Azorhizobium* and *Mesorhizobium*) of the alpha (α) subclass of Proteobacteria have been classified, mainly by comparison of the sequences of 16S rRNA genes. Some species of *Sinorhizobium* were transferred to the genus *Ensifer*. The genera *Rhizobium*, *Bradyrhizobium*, *Sinorhizobium* and *Mesorhizobium* were recognized mainly on root nodule formation. While *Azorhizobium* was firstly established for the ability of the first species, *Azo. caulinodans*, to form nodules on both stems and roots of *Ses. rostrata* [13]. It was later found that stem nodulation is not restricted to *Azorhizobium*. The following species, *Azo. doebereinerae*, was not able to form stem nodules with *Ses. rostrata*, but induced pseudonodules instead. Conversely, *Azo. caulinodans* was not able to complete nodule formation in *Sesbania virgata*, although pseudonodules were observed [19]. *Ensifer saheli* (formerly *Sinorhizobium saheli*) and *Ensifer terangae* (formerly *Sinorhizobium terangae*) biovar sesbaniae strains were found to effectively stem nodulate *Ses. rostrata*. However, clear differences in symbiotic aspects between *Ensifer* and *Azorhizobium* were found. *Ens. saheli* and *Ens. terangae* biovar sesbaniae were effective symbionts with all *Sesbania* species tested, while *Azorhizobium* strains fixed nitrogen only in symbiosis with *Ses. rostrata*. Both species of *Ensifer* were incapable of asymbiotic nitrogenase activity. Among *Sesbania* symbionts, *Azorhizobium* can easily be distinguished from *Ensifer* on the basis of symbiotic and free-living nitrogen fixation. Even though stem nodulation is not restricted to *Azorhizobium*, it was found that *Azorhizobium* was highly specific for stem nodulation of *Ses. rostrata* [20]. Some *Rhizobium* strains were also found to produce stem and root nodules on *Ses. rostrata* but their stem nodules exhibited very low activity or were ineffective [21]. Photosynthetic strains of *Bradyrhizobium* specifically nodulated stems of some leguminous species of the genus *Aeschynomene* such as *Aes. sensitiva* and *Aes. indica*. Evidence that bacterial photosynthesis plays a role in the efficiency of stem nodulation was also provided [22].

Besides 5 genera originally proposed as rhizobial genera, the other soil bacteria of the alpha subclass of Proteobacteria have been reported as nitrogen-fixing symbionts of legumes. These bacteria include *Methylobacterium nodulans* isolated from a herbal legume, *Crotalaria* sp. [23], *Methylobacterium* spp. [24], *Devosia neptuniae* isolated from an aquatic legume, *Nep. natans* [25], *Phyllobacterium trifolii* isolated from *Trifolium pratense* (red clover) [26], *Phyllobacterium* sp. [27], *Phyllobacterium ifriqiyense*, *Phyllobacterium leguminum* [28], *Ochrobactrum* sp. isolated from *Acacia mangium* [29], *Ochrobactrum lupini* isolated from *Lupinus* spp. [30], *Ochrobactrum cytisi* isolated from *Cytisus scoparius* [31], the unidentified microsymbiont that is closely related to 3 genera; *Balneomonas, Bosea* and *Chelatococcus* [32], *Shinella kummerowiae* isolated from a herbal legume, *Kummerowia stipulacea* [33] and 3 species of *Microvirga* [34]. Even though *Shi. kummerowiae* did not form nodules with its original host, *Kum. stipulacea*, as well as the tested hosts including *Glycine max* (soybean), *Leucaena leucocephala* (white popinac), *Medicago sativa* (alfalfa) and *Phaseolus vulgaris* (common bean), it has been proposed as a symbiotic bacterium as it contains the symbiotic genes *nifH*, *nodC* and *nodD* [33].

Rhizobial members have also been defined within the beta (β) subclass of Proteo-bacteria. *Cupriavidus taiwanensis* (formerly *Ralstonia taiwanensis*), *Cupriavidus* spp. and *Ralstonia* spp. have been reported as the legume symbionts [27, 35-44]. Several species of *Burkholderia* were found to nodulate several legumes [27, 35, 37, 38, 41, 43, 45-54]. *Herbaspirillum lusitanum* was found to nodulate *Pha. vulgaris* [55]. *Pseudomonas* sp. has been discovered as the first member of the gamma (γ) subclass of Proteobacteria that nodulate *Robinia pseudoacacia* (black locust) [52]. The classification and taxonomy of rhizobia are shown in Table **1**.

1.4. SYMBIOTIC GENE TRANSFER AMONG RHIZOBIAL MEMBERS

Rhizobial members in taxonomically diverse genera indicate the transfer of symbiotic genes. Phylogenetic analyses of the symbiotic genes *nifH* and *nodA* from the alpha and beta subclasses of Proteobacteria suggest that rhizobial members in the beta subclass have evolved from diazotrophs through multiple lateral gene transfers [38]. In some genera, such as *Cupriavidus*, *Ensifer/Sinorhizobium*, *Mesorhizobium* and *Rhizobium*, genes essential for nitrogen fixation and nodulation are located on megaplasmids termed as "symbiotic plasmids (pSym)". In some genera, such as *Bradyrhizobium* and *Mesorhizobium*, symbiotic genes are located on chromosome. This DNA segment is termed as "symbiosis island". Some species of *Mesorhizobium* possess pSym, while some species of *Mesorhizobium* possess symbiosis island [56]. Since pSym and symbiosis island can be transferred to other bacteria, some saprophytic or rhizospheric bacteria may become symbiotic by the horizontal acquisition of symbiotic genes [57]. The population level analyses of *Rhizobium* species also revealed the existence of lateral transfer of pSym within species in agricultural fields and pastures [58-61]. Horizontal transfers of pSym and symbiosis island have been found to play a crucial role in rhizobial evolution. Ample evidence is given by the experimental evolution of a pathogenic *Ralstonia solanacearum* that received pSym of *Cup. taiwanensis* and became a *Mimosa*-nodulating symbiont [62]. *Ensifer adhaerens* is a soil bacterium that attaches to other bacteria and may cause lysis of the other bacteria. Based on the sequence of 16S rRNA gene, *Ens. adhaerens* is related to *Sinorhizobium* spp. *Ens. adhaerens* ATCC 33499 did not nodulate *Leu. leucocephala* or *Pha. vulgaris*, but with symbiotic plasmids from *Rhizobium tropici* CFN299, it formed nitrogen-fixing nodules on both hosts [63].

Natural selection of adaptive changes in the legume environment following horizontal transfer has been a major driving force in rhizobial evolution and diversification that show the potential of experimental evolution to decipher the mechanisms leading to symbiosis [62].

1.5. NON-SYMBIOTIC RHIZOBIA

Even though symbiotic rhizobia belonging to a wide range of bacterial genera have been reported, the previous studies have reported a major population of non-symbiotic rhizobia in different species such as *Bradyrhizobium elkanii, Bradyrhizobium japonicum* [64], *Rhizobium etli* [65] and *Rhizobium leguminosarum* [65-67]. Non-symbiotic rhizobial isolates that were related to *Mes. loti* (formerly *Rhizobium loti*) were recovered from the rhizosphere of *Lotus corniculatus* [68]. Non-symbiotic isolates of *Rhi. leguminosarum* and *Rhi. etli* were more numerous than symbiotic isolates in rhizospheres and soil samples [65-67]. Symbiotic properties in some rhizobia are genetically unstable, raising the possibility that non-symbiotic rhizobia are a significant component of rhizobial populations in the soil [68].

1.6. SYMBIOTIC PROMISCUITY OF RHIZOBIA

Although the host specificity is considered as an important feature of rhizobia-legume symbioses, it seems a false impression as many symbionts have been found to intimately associate with many different partners. The broad range of this association is termed as "symbiotic promiscuity". It has been frequently found that the same host plants can be nodulated by bacteria in different genera or species. Pongsilp *et al.* [69] suggest that bacteria of different genera may adapt to the environmental conditions influenced by root exudates from their hosts. As root exudates are composed of both low and high molecular weight components, including an array of primary and secondary metabolites, proteins and peptides [70, 71] that vary in quantity and chemical structure depending on the plant species, therefore root exudates can provide the selective environments for specific groups of bacteria. The presence of rhizobia which belong to different genera in the same host nodule might be a result of genetic diversification and adaptation of the bacteria to their environments [72]. Several hosts such as *Leu.*

leucocephala, Macroptilium atropupureum (siratro), *Med. sativa, Pha. vulgaris* and *Vigna unguiculata* (cowpea) are well known to be highly promiscuous. As shown in Table **1**, many leguminous plants can be nodulated by more than one rhizobial species. In pararell, it has been also found that the nodulation of host plants in different genera is caused by the same rhizobial species or strains. *Ensifer fredii* (formerly *Rhizobium* sp.) NGR234 exhibited its exceptionally broad host range which includes 232 legume species (distributed in 112 genera) and formed ineffective nodules with the non-legume *Parasponia andersonii*. *Ens. fredii* (formerly *Rhizobium fredii*) USDA 257 effectively and ineffectively nodulated 135 legume species (distributed in 79 genera). The most striking correlation is that all 135 legume species nodulated by *Ens. fredii* USDA 257 are also host to *Ens. fredii* NGR234. Thus, with respect to nodulation, the host range of USDA 257 is a subset nested entirely within that of NGR234 [73]. As also shown in Table **1**, almost rhizobial species are determined to be promiscuous as they are able to nodulate more than one legume species.

1.7. RHIZOBIAL DIVERSITY

Rhizobial diversity has been revealed by many studies and almost all of the data reported previously indicates a high level of diversity. The diversity is assessed in terms of "phenotypic and genotypic diversity". Phenotypic diversity can be characterized by several methods such as numerical analysis based on phenotypic characteristics, enzyme patterns, serological study, composition of extracellular polysaccharides, composition of fatty acids, bacteriophage typing data, soluble protein patterns and symbiotic efficiencies. Genotypic diversity can be characterized by the techniques based on DNA analyses. Molecular techniques used frequently are i) sequence analysis; ii) polymerase chain reaction (PCR) fingerprinting; iii) amplified fragment length polymorphism (AFLP); iv) restriction fragment length polymorphism (RFLP); v) PCR-RFLP and vi) plasmid profile. These techniques are powerful tools for revealing the genetic diversity and phylogeny of bacteria. These techniques provide a different level of perspective to interpret the phenotypic and genotypic variations among legume symbionts. Most studies employed the combination of several methodologies to characterize and to examine genetic relationships of these specific groups of bacteria. The ranges of discriminating power, respective levels of resolution and limitations have been evaluated. The application of these techniques relies upon many factors such as desired discriminative power, sensitivity of methods, precision of results, specificity of the regions, amount of DNA used, fragment of interests, sequence knowledge, complexity of the populations, procedures, instruments, specialized software packages, time and labor consumption.

An assessment of the genetic diversity and genetic relationships among strains could provide valuable information about bacterial genotypes that are well adapted to a certain environment [74]. Considerable genetic diversity was observed in many groups of rhizobial populations such as i) *Rhi. leguminosarum* and *Ensifer meliloti* (formerly *Rhizobium meliloti*) strains [75]; ii) non-symbiotic *Rhi. leguminosarum* strains in Mexico [65]; iii) rhizobial isolates obtained from nodules of *Cicer arietinum* (chickpea) growing in uninoculated fields over a wide geographic range [76]; iv) *Rhi. leguminosarum* biovar viciae isolated directly from bulk soil [59]; v) *Ens. meliloti* populations from nodules of different legumes [77]; vi) rhizobial isolates from nodules of *Pha. vulgaris* growing in northwestern Argentina [78]; vii) rhizobial populations nodulating native shrubby legumes in open eucalypt forest of southeastern Australia [79]; viii) rhizobial isolates nodulating 4 *Acacia* species from different sites in Morocco [80]; ix) rhizobial isolates nodulating *Medicago ruthenica* native to Inner Mongolia [81]; x) rhizobial isolates from root nodules of *Mimosa affinis* in Mexico [82]; xi) rhizobial isolates from root nodules of *Amorpha fruticosa* (false indigo) [83]; xii) *Med. sativa*-nodulating rhizobia isolated from acid soils of Argentina and Uruguay [84]; xiii) *Ens. meliloti* (formerly *Sinorhizobium meliloti*) populations from different plant species [85]; xiv) rhizobial isolates obtained from root nodules of *Pha. vulgaris* cultivated in soils originating from different agroecological areas in Senegal and Gambia [86]; xv) *Bradyrhizobium* populations isolated directly from different soil samples in Thailand [64]; xvi) rhizobial isolates nodulating 15 *Lespedeza* spp. in northern hemisphere in the United States and China [87]; xvii) rhizobial isolates from root nodules of *Indigofera* and *Kummerowia*, growing in the Loess plateau of China [88]; xviii) rhizobial isolates from root nodules of *Astragalus* and *Lespedeza* spp. growing in the Loess plateau of China [89]; xix) *Rhizobium galegae* strains isolated from root nodules of wild *Galega orientalis* and *Galega officinalis* in the Caucasus of Russia [90]; xx) rhizobial populations nodulating *Pha. vulgaris* cultivated in a traditionally managed Milpa plot in Mexico [91]; xxi) rhizobial populations isolated from root nodules of 18 agroforestry species growing in diverse ecoclimatic zones in southern Ethiopia [24]; xxii) rhizobial populations nodulating 3 medicinal legumes, including *Derris elliptica*, *Indigofera tinctoria* (true indigo) and *Pueraria mirifica* (white Kwao Kruea), growing in 16 provinces of Thailand [44]; xxiii) rhizobial isolates from root nodules of *Med. sativa* and *Melilotus alba* (sweet clover) growing in Canada [92].

Table 1: Classification and taxonomy of rhizobia

Class	Family	Genus	Species	Host	Reference
Alpha-Proteo bacteria	Rhizobiaceae	*Rhizobium*	*Rhizobium alkalisoli*	*Caragana intermedia* (*Caragana korshinskii*)	[93]
				Caragana microphylla	[93]
				Phaseolus vulgaris	[93]
				Vigna radiata	[93]
			Rhizobium cellulosilyticum	*Glycyrrhiza glabra* (ineffective)	[94]
				Medicago sativa (ineffective)	[95]
			Rhizobium daejeonense	*Med. sativa*	[96]
			Rhizobium fabae	*Vicia faba*	[97]
			Rhizobium etli (formerly *Rhizobium leguminosarum* biovar phaseoli type I) [98]	*Acacia senegal*	[86]
				Acacia seyal	[86]
				Faidherbia albida	[86]
				Leucaena leucocephala	[82, 86]
				Macroptilium atropurpureum (ineffective)	[91]
				Mimosa affinis	[82]
				Pha. vulgaris	[82, 86, 91, 98-100]
				Vic. faba	[101]
				Vig. radiata	[102]
				Vigna unguiculata	[102, 103]
			Rhizobium galegae	*Galega officinalis* (ineffective/effective)	[104, 105]
				Galega orientalis (ineffective/effective)	[100, 104, 105]
				Glycyrrhiza glabra	[94]
				Glycyrrhiza uralensis	[94]
				Leu. leucocephala	[106]

Table 1: cont….

					Sesbania herbacea	[106]
					Sesbania rostrata (ineffective)	[106]
					Trifolium repens (ineffective)	[106]
				Rhizobium gallicum	Glycyrrhiza glabra	[94]
					Glycyrrhiza uralensis	[94]
					Leu. leucocephala	[91]
					Macroptilium atropurpureum	[91]
					Pha. vulgaris	[91, 99, 107]
				Rhizobium giardinii	Acacia abyssinica	[24]
					Glycyrrhiza glabra (ineffective)	[94]
					Glycyrrhiza uralensis (ineffective)	[94]
					Pha. vulgaris	[99, 107]
				Rhizobium grahamii	Clitoria ternatea	[108]
					Dalea leporina	[108]
					Leu. leucocephala	[108]
				Rhizobium hainanense	Desmodium sinuatum	[109]
				Rhizobium herbae	Astragalus membranaceus	[110]
				Rhizobium huautlense	Leu. leucocephala	[106]
					Pha. vulgaris	[24]
					Ses. herbacea	[106]
					Sesbania sesban	[24]
					Vig. unguiculata	[24]
				Rhizobium indigoferae	Indigofera amblyantha	[88]
					Indigofera carlesii	[88]
					Indigofera pataninii	[88]
				Rhizobium leguminosarum (synonym *Rhizobium trifolii*) [111]	Acacia confusa	[112]
					Acacia farnesiana (Vachellia farnesiana)	[112]
					Albizia procera	[112]
					Cajanus cajan	[112]
					Cicer arietinum	[113]
					Gliricidia sepium	[112]
					Glycyrrhiza glabra (ineffective)	[94]
					Glycyrrhiza uralensis	[94]
					Lathyrus japonicus	[112]
					Lens esculenta (Lens culinaris)	[99]

Table 1: cont….

				Leu. leucocephala	[112]
				Mimosa invisa	[112]
				Mimosa pudica	[112]
				Parasponia andersonii	[114]
				Pha. vulgaris	[99, 112, 115, 116]
				Pisum sativum	[99, 100, 112, 115, 117, 118]
				Trifolium dubium	[112]
				Trifolium pratense	[99, 112, 116]
				Trifolium repens	[100, 112]
				Trifolium subterraneum	[100, 116, 119]
				Samanea saman	[112]
				Sesbania speciosa	[112]
				Vicia cracca	[112, 117]
				Vic. faba	[101, 112, 116, 118]
				Vicia hirsuta	[112, 120]
				Vicia sativa	[112]
				Vicia tetrasperma	[112]
				Vicia unijuga	[112]
				Vig. radiata	[102]
				Vig. unguiculata	[102]
			Rhizobium leucaenae (formerly *Rhizobium tropici* type A) [121]	*Gli. sepium*	[121]
				Leucaena esculenta	[121]
				Leu. leucocephala	[121]
				Pha. vulgaris	[121]
			Rhizobium loessense	*Astragalus adsurgens*	[89]
				Astragalus chrysopterus	[89]
				Astragalus complanatus	[89]
				Astragalus scobwerrimus	[89]
			Rhizobium lupini	*Lupinus albus*	[122]
			Rhizobium lusitanum	*Pha. vulgaris*	[123]
			Rhizobium mesoamericanum	*Macroptilium atropurpureum*	[108]
				Mim. pudica	[108]
				Pha. vulgaris	[108]

Table 1: cont….

				Vig. unguiculata	[108]
			Rhizobium mesosinicum	*Albizia julibrissin*	[124]
			Rhizobium miluonense	*Lespedeza chinensis*	[125]
				Pha. vulgaris (ineffective)	[125]
			Rhizobium mongolense	*Leu. leucocephala*	[100]
				Medicago rustica	[81]
				Medicago ruthenica	[81]
			Rhizobium multihospitium	*Halimodendron halodendron*	[126]
				Robinia pseudoacacia	[126]
			Rhizobium phaseoli	*Pha. vulgaris*	[127]
			Rhizobium pisi (formerly *Rhizobium leguminosarum*) [111]	*Pis. sativum*	[111]
			Rhizobium pusense	*Cicer arietinum*	[128]
			Rhizobium sullae (formerly *Rhizobium hedysari*) [129]	*Hedysarum coronarium*	[129]
			Rhizobium tibeticum	*Medicago lupulina*	[130]
				Med. sativa	[130]
				Melilotus officinalis	[130]
				Pha. vulgaris	[130]
				Trigonella archiducisnicolai	[130]
				Trigonella foenum-graecum	[130]
			Rhizobium tropici (formerly *Rhizobium leguminosarum* biovar phaseoli type II) [116]	*Acacia senegal*	[86]
				Acacia seyal	[86]
				Amorpha fruticosa	[83]
				Fai. albida	[86]
				Leu. esculenta	[116]
				Leu. leucocephla	[86, 99, 116]
				Leucaena spp.	[131]
				Macroptilium atropurpureum	[131]
				Pha. vulgaris	[86, 99, 100, 115, 116, 123, 131, 132]
			Rhizobium tubonense	*Med. sativa*	[133]
				Vig. unguiculata	[133]

Table 1: cont….

			Rhizobium undicola (formerly *Allorhizobium undicola*) [15]	*Acacia seyal* (ineffective)	[14]
				Acacia tortilis (ineffective)	[14]
				Fai. albida (ineffective)	[14]
				Lotus arabicus (ineffective)	[14]
				Med. sativa (ineffective)	[14]
				Neptunia natans	[14]
			Rhizobium vallis	*Indigofera spicata*	[134]
				Mim. pudica	[134]
				Pha. vulgaris	[134]
			Rhizobium vignae	*Astragalus dahuricus* (*Astragalus mongholicus*)	[135]
			Rhizobium yanglingense	*Pha. vulgaris* (inefffective)	[136]
			Rhizobium spp.	*Acacia albida*	[29]
				Acacia cyanophylla	[137]
				Acacia farnesiana	[138]
				Acacia mangium	[29]
				Acacia tortilis	[24]
				Acacia sp.	[139]
				Amo. fruticosa (ineffective)	[83]
				Ast. adsurgens	[89]
				Ast. complanatus	[89]
				Astragalus spp.	[100]
				Caj. cajan	[138]
				Calliandra calothyrsus	[140]
				Cic. arietinum	[112]
				Clitoria sp.	[141]
				Colutea arborescens	[141]
				Dalbergia sp.	[27]
				Derris elliptica	[44, 142]
				Desmanthus illinoensis	[141]
				Desmanthus virgatus	[140]
				Entada phaseoloides	[112]
				Gli. sepium	[138, 140]
				Hedysarum scoparium	[89]
				Indigofera tinctoria	[42, 44]
				Laburnum anagyroides	[141]
				Lathyrus hirsutus	[138]

Table 1: cont....

				Lens culinaris	[138]
				Lespedeza cyrtobotrya	[89]
				Lespedeza davidii	[89]
				Leu. leucocephala (ineffective/effective)	[100, 103, 116, 138, 140]
				Lotus tenuis	[143]
				Med. sativa (pseudonodules/ ineffective/effective)	[92, 138, 140, 144]
				Melilotus alba (Melilotus albus)	[92]
				Millettia ferruginea	[24]
				Mimosa diplotricha	[38]
				Mim. invisa	[112]
				Mimosa pigra	[35]
				Mim. pudica	[35, 38]
				Onobrychis vicifolia	[141]
				Onobrychis spp.	[100]
				Oxytropis spp.	[100]
				Pachyrhizus erosus	[72]
				Paraserianthes falcataria (Falcataria moluccana)	[29]
				Pha. vulgaris (ineffective/effective)	[92, 103, 116, 138, 140]
				Prosopis spp.	[140]
				Pterocarpus klemmei	[112]
				Pueraria mirifica	[44, 69]
				Pueraria phaseoloides	[141]
				Sesbania aculeata (Sesbania bispinosa)	[112]
				Sesbania aegyptiaca	[112]
				Sesbania grandiflora	[138]
				Sesbania macrocarpa	[138]
				Ses. rostrata	[112, 138]
				Ses. sesban	[112]
				Sophora chrysophylla	[140]
		Ensifer/ Sinorhizobium	Sinorhizobium abri	Abrus precatorius	[145]
			Sinorhizobium americanus	Acacia spp.	[146]
			Ensifer arboris (formerly Sinorhizobium arboris) [147]	Acacia senegal	[148]
				Prosopis chilensis	[148]

Table 1: cont….

				Sinorhizobium chiapanecum	*Acaciella angustissima*	[149]
				Ensifer fredii (formerly *Sinorhizobium fredii*) [147] and *Rhizobium fredii* [10]) (*Rhizobium* sp. NGR234 is transferred to *Sinorhizobium fredii* NGR234) [150]	*Acacia auriculiformis* (ineffective/effective)	[73]
					Acacia bonariensis (ineffective)	[73]
					Acacia cyanophylla (ineffective)	[73]
					Acacia farnesiana (ineffective)	[73]
					Acacia macracantha (*Vachellia macracantha*) (ineffective)	[73]
					Acacia mangium (ineffective)	[73]
					Acacia mearnsii (ineffective)	[73]
					Acacia pendula	[73]
					Acacia retinodes (ineffective/effective)	[73]
					Acacia saligna (ineffective/effective)	[73]
					Aeschynomene aspera (ineffective)	[73]
					Aeschynomene falcata (ineffective)	[73]
					Aeschynomene indica (ineffective)	[73]
					Alb. julibrissin (ineffective)	[73]
					Albizia lebbeck	[73]
					Alb. procera	[73]
					Albizia saponaria (ineffective)	[73]
					Alysicarpus vaginalis	[73]
					Amo. fruticosa	[73]
					Amphicarpaea trisperma	[73]
					Anagyris foetida	[73]
					Anthyllis vulneraria (ineffective)	[73]
					Aotus ericoides (ineffective/effective)	[73]
					Apios americana (ineffective/effective)	[73]
					Arachis hypogaea (ineffective)	[73]
					Ateleia ovata	[73]
					Bolusanthus speciosus (ineffective)	[73]
					Caj. cajan	[73]

Table 1: cont....

				Cajanus scarabaeoides	[73]
				Calicotome villosa (ineffective)	[73]
				Calliandra houstoniana (ineffective)	[73]
				Calopogonium caeruleum	[73]
				Calopogonium mucunoides (ineffective)	[73]
				Canavalia ensiformis (ineffective)	[73]
				Canavalia rosea (ineffective/effective)	[73]
				Centrosema pubescens (ineffective)	[73]
				Chamaecrista fasciculata (ineffective/effective)	[73]
				Chamaecytisus proliferus (ineffective)	[73]
				Chorizema dicksonii (ineffective)	[73]
				Chorizema diversifolium	[73]
				Codariocalyx gyroides	[73]
				Codariocalyx motorius	[73]
				Col. arborescens	[73]
				Cratylia argentea (ineffective)	[73]
				Crotalaria juncea (ineffective)	[73]
				Crotalaria sericea (ineffective/effective)	[73]
				Cytisus hirsutus (ineffective)	[73]
				Cytisus villosus (ineffective)	[73]
				Dalbergia martini (ineffective)	[73]
				Dalbergia retusa	[73]
				Dalea candida	[73]
				Dalea purpurea	[73]
				Delonix regia (ineffective)	[73]
				Dendrolobium triangulare	[73]
				Desmanthus illinoensis (ineffective/effective)	[73]
				Desmanthus virgatus	[73]
				Desmodium canadense	[73]
				Desmodium dichotomum	[73]
				Desmodium intortum	[73]
				Desmodium uncinatum	[73]

Table 1: cont….

				Dicerma biarticulatum (ineffective)	[73]
				Dichrostachys cinerea	[73]
				Dillwynia glaberrima (ineffective)	[73]
				Dioclea guianensis (ineffective)	[73]
				Dioclea sericea (ineffective)	[73]
				Dioclea virgata (ineffective)	[73]
				Dolichos junghuhnianus	[73]
				Dolichos trilobus	[73]
				Dunbaria circinalis	[73]
				Dunbaria nivea	[73]
				Dunbaria villosa (ineffective/effective)	[73]
				Dysolobium apioides (ineffective/effective)	[73]
				Enterolobium contortisiliquum (ineffective/effective)	[73]
				Enterolobium timbouva	[73]
				Eriosema violaceum (ineffective)	[73]
				Erythrina abyssinica	[73]
				Erythrina costaricensis	[73]
				Erythrina cristagalli	[73]
				Erythrina fusca	[73]
				Erythrina variegata (ineffective)	[73]
				Erythrina vespertilio	[73]
				Fai. albida	[73]
				Flemingia congesta	[73]
				Flemingia strobilifera	[73]
				Galactia jussiaeana (ineffective/effective)	[73]
				Galactia latisiliqua (ineffective)	[73]
				Galactia striata (ineffective)	[73]
				Gastrolobium bilobum (ineffective)	[73]
				Glycine canescens	[73]
				Glycine max	[73, 100, 112, 151-156]
				Glycine soja (ineffective/effective)	[73, 153]

Table 1: cont….

					Glycine tabacina	[73]
					Glycine tomentella	[73]
					Glycyrrhiza glabra	[73]
					Gompholobium latifolium (ineffective)	[73]
					Goodia lotifolia (ineffective/effective)	[73]
					Gueldenstaedtia stenophylla (ineffective)	[73]
					Hal. halodendron (ineffective)	[73]
					Hardenbergia comptoniana	[73]
					Hardenbergia violacea (ineffective/effective)	[73]
					Hedysarum alpinum (ineffective)	[73]
					Hesperolaburnum platycarpum	[73]
					Hovea acutifolia (ineffective)	[73]
					Hovea linearis (ineffective)	[73]
					Indigofera arrecta	[73]
					Indifogera australis (ineffective)	[73]
					Indigofera jamaicensis	[73]
					Ind. tinctoria	[73]
					Inga mortoniana	[73]
					Kennedia beckxiana (ineffective)	[73]
					Kennedia nigricans (ineffective/effective)	[73]
					Kennedia prostrata (ineffective/effective)	[73]
					Kennedia rubicunda (ineffective/effective)	[73]
					Kummerowia stipulacea	[73]
					Kummerowia striata	[73]
					Lablab purpureus	[73]
					Laburnum anagyroides (ineffective)	[73]
					Laburnum vossii (ineffective)	[73]
					Lembotropis nigricans (ineffective)	[73]
					Lespedeza bicolor	[73]
					Leucaena diversifolia (ineffective)	[73]
					Leu. leucocephala	[73]

Table 1: cont….

				Leucaena trichodes (ineffective)	[73]
				Lotus corniculatus (ineffective/effective)	[73]
				Lotus halophilus (ineffective)	[73]
				Lotus japonicus	[73]
				Lotus pedunculatus (ineffective)	[73]
				Lupinus pilosus (ineffective)	[73]
				Macroptilium atropurpureum	[73]
				Macroptilium bracteatum	[73]
				Macroptilium lathyroides	[73]
				Macroptilium longepedunculatum (ineffective)	[73]
				Macrotyloma axillare	[73]
				Macrotyloma uniflorum	[73]
				Medicago cancellata (ineffective)	[73]
				Medicago hispida	[112]
				Medicago papillosa (ineffective)	[73]
				Med. sativa	[112]
				Millettia megasperma (ineffective)	[73]
				Mirbelia dilatata (ineffective)	[73]
				Mirbelia pungens	[73]
				Mundulea sericea	[73]
				Neonotonia wightii	[73]
				Nep. natans (ineffective)	[73]
				Otoptera burchellii	[73]
				Oxylobium ellipticum (ineffective)	[73]
				Oxytropis halleri	[73]
				Pachecoa prismatica (ineffective)	[73]
				Pachyrhizus erosus (ineffective)	[73]
				Pachyrhizus tuberosus	[73]
				Paraserianthes falcataria	[73]
				Parasponia andersonii (non-legume) (ineffective)	[73]
				Phaseolus acutifolius (ineffective)	[73]
				Phaseolus angustifolius (ineffective)	[73]

Table 1: cont….

					Phaseolus coccineus (ineffective/effective)	[73]
					Phaseolus leptostachyus	[73]
					Phaseolus polystachyus (*Phaseolus polystachios*)	[73]
					Pha. vulgaris (ineffective/effective)	[73, 157, 158]
					Phyllodium elegans (ineffective)	[73]
					Piptanthus concolor	[73]
					Pseudarthria viscida (ineffective)	[73]
					Psophocarpus palustris	[73]
					Psophocarpus tetragonolobus	[73]
					Psoralea plumose	[73]
					Psoralea pustulata (ineffective/effective)	[73]
					Pterocarpus lucens (ineffective/effective)	[73]
					Pueraria lobata (*Pueraria montana*) (ineffective)	[73]
					Pue. phaseoloides (ineffective)	[73]
					Pultenaea blakelyi	[73]
					Pultenaea daphnoides (ineffective)	[73]
					Pultenaea microphylla (ineffective)	[73]
					Pultenaea villosa (ineffective)	[73]
					Pycnospora lutescens	[73]
					Retama monosperma	[73]
					Rhynchosia minima	[73]
					Rhynchosia rothii	[73]
					Rhynchosia sublobata	[73]
					Robinia hispida (ineffective/effective)	[73]
					Rob. pseudoacacia	[73]
					Sam. saman	[73]
					Sesbania bispinosa (ineffective)	[73]
					Sesbania cannabina (ineffective)	[73]
					Ses. grandiflora	[73]
					Ses. rostrata (ineffective)	[73]
					Sophora davidii	[73]

Table 1: cont….

				Sophora tomentosa	[73]
				Sophora velutina	[73]
				Spartium junceum (ineffective)	[73]
				Stylosanthes guianensis (ineffective)	[73]
				Stylosanthes hamata (ineffective)	[73]
				Stylosanthes humilis (ineffective)	[73]
				Stylosanthes scabra (ineffective)	[73]
				Swainsona forrestii	[73]
				Templetonia retusa (ineffective)	[73]
				Tephrosia cinerea	[73]
				Tephrosia rosea	[73]
				Tephrosia sessiliflora	[73]
				Tephrosia vogelii	[73]
				Teramnus labialis	[73]
				Teramnus uncinatus (ineffective/effective)	[73]
				Teyleria koordersii (ineffective)	[73]
				Tipuana tipu (ineffective)	[73]
				Ulex europaeus (ineffective)	[73]
				Vigna aconitifolia	[73]
				Vigna angularis	[73]
				Vigna caracalla (*Cochliasanthus caracalla*)	[73]
				Vigna cylindrical	[73]
				Vigna glabrescens (ineffective/effective)	[73]
				Vigna hosei	[73]
				Vigna lanceolata	[73]
				Vigna luteola	[73]
				Vigna minima	[73]
				Vigna mungo	[73]
				Vigna oblongifolia	[73]
				Vigna parkeri	[73]
				Vig. radiata (ineffective/effective)	[73, 102]
				Vigna subterranean	[73]
				Vigna trilobata	[73]
				Vigna umbellata	[73]

Table 1: cont….

				Vig. unguiculata	[73, 102]
				Vigna vexillata	[73]
				Virgilia capensis	[73]
				Virgilia divaricata (ineffective)	[73]
				Wisteria frutescens (ineffective)	[73]
				Wisteria sinensis (ineffective)	[73]
				Xeroderris stuhlmannii	[73]
				Zornia diphylla (ineffective)	[73]
				Zornia latifolia (ineffective)	[73]
			Ensifer garamanticus	Argyrolobium uniflorum	[159]
				Lot. arabicus	[159]
				Lotus creticus	[159]
				Med. sativa	[159]
			Sinorhizobium indiaensis	Ses. rostrata	[145]
			Ensifer kostiensis (formerly Sinorhiobium kostiensis) [147]	Acacia senegal	[148]
				Pro. chilensis	[148]
			Ensifer kummerowiae (formerly Sinorhizobium kummerowiae) [147]	Kum. stipulacea	[88]
				Med. sativa	[88]
			Ensifer medicae (formerly Sinorhizobium medicae) [147]	Medicago orbicularis	[100, 160]
				Medicago polymorpha	[92, 160]
				Medicago rugosa	[160]
				Med. sativa	[92]
				Medicago truncatula	[100]
				Medicago trunculata	[160]
				Mel. alba	[92]
				Pha. vulgaris (ineffective)	[92]
			Ensifer meliloti (formerly Sinorhizobium meliloti [147] and Rhizobium meliloti [161])	Chamaecrista fasciculata	[73]
				Glycine max	[112]
				Glycyrrhiza glabra (ineffective/effective)	[94]
				Glycyrrhiza uralebsis (ineffective/effective)	[94]
				Lespedeza spp.	[162]
				Med. hispida	[112]

Table 1: cont….

				Med. polymorpha (ineffective)	[160]
				Med. sativa	[84, 92, 99, 100, 112, 140, 163, 164]
				Med. truncatula	[165]
				Mel. alba	[92]
				Pha. vulgaris (pseudonodules/ineffective)	[92, 140]
				Vic. faba	[101]
			Ensifer mexicanus	*Acacia angustissima*	[166]
			Ensifer numidicus	*Arg. uniflorum*	[159]
				Lot. arabicus	[159]
				Lot. creticus	[159]
				Med. sativa	[159]
			Ensifer saheli (formerly *Rhizobium saheli*) [147]	*Les. cyrtobotrya*	[87]
				Lespedeza daurica	[87]
				Lespedeza inschanica	[87]
				Lespedeza tomentosa	[87]
				Ses. canabina	[100, 161]
				Sesbania grandiflora	[161]
				Sesbania pachycarpa (*Sesbania brachycarpa*)	[161]
				Ses. rostrata	[20, 167]
			Ensifer sojae	*Glycine max*	[168]
				Glycine soja	[168]
				Vig. unguiculata	[168]
			Ensifer terangae (formerly *Sinorhizobium terangae*) [147]	*Acacia laeta*	[100, 161]
				Acacia spp.	[24, 167]
				Ses. rostrata	[20, 161]
				Sesbania spp.	[24]
			Ensifer xinjiangense (formerly *Sinorhizobium xinjiangense*) [147]	*Glycine max*	[169]
			Ensifer spp. and *Sinorhizobium* spp.	*Acacia cyanophylla*	[80]
				Acacia gummifera	[80]
				Acacia horrida	[80]
				Acacia raddiana	[80]
				Acacia spp.	[24]

Table 1: cont….

				Der. elliptica	[44, 142]
				Ind. tinctoria	[42, 44]
				Macroptilium atropurpureum (ineffective)	[92]
				Med. lupulina (ineffective)	[92]
				Med. sativa (ineffective)	[92]
				Mel. alba (ineffective)	[92]
				Mim. pudica	[38]
				Parasponia spp.	[170]
				Pha. vulgaris (ineffective)	[92]
		Shinella	*Shinella kummerowiae*	*Kum. stipulacea* (nodule endophyte)	[33]
	Bradyrhizo-biaceae	*Bradyrhizobium*	*Bradyrhizobium canariense*	*Adenocarpus* spp.	[171]
				Chamaecytisus proliferus	[171]
				Lupinus luteus	[171]
				Macroptilium atropurpureum	[171]
				Ornithopus spp.	[171]
				Spartocytisus supranubius (*Cytisus supranubius*)	[171]
				Teline stenopetala (*Genista stenopetala*)	[171]
			Bradyrhizobium cytisi	*Cyt. villosus*	[172]
			Bradyrhizobium denitrificans (formerly *Blastobacter denitrificans*) [18]	*Aes. indica*	[17, 18]
			Bradyrhizobium elkanii (formerly *Bradyrhizobium japonicum*) [173]	*Astragalus* spp.	[100]
				Caj. cajan	[174]
				Fai. albida	[24]
				Glycine max	[100, 173]
				Ind. tinctoria	[42, 44]
				Les. bicolor	[87]
				Lespedeza capitata	[87]
				Lespedeza cuneata	[87]
				Lespedeza juncea	[87]
				Lespedeza procumbens	[87]
				Lespedeza stipulacea	[87]
				Lespedeza striata	[87]
				Lespedeza spp.	[162]

Table 1: cont….

				Macroptilium atropurpureum	[41]
				Mil. ferruginea	[24]
				Rhynchosia caribaea	[41]
				Rhynchosia ferulifolia	[41]
				Rhy. minima	[41]
				Vig. radiata	[102]
				Vig. unguiculata	[102]
			Bradyrhizobium japonicum (formerly *Rhizobium japonicum*) [175]	*Acacia saligna*	[24]
				Aeschynomene americana	[112]
				Aes. falcata	[112]
				Albizia gummifera	[24]
				Alb. julibrissin	[112]
				Amphicarpaea edgeworthii	[112]
				Ara. hypogaea	[112]
				Astragalus spp.	[100]
				Caj. cajan	[24, 112]
				Calopogonium caeruleum	[112]
				Calopogonium mucunoides	[112]
				Cassia nomame	[112]
				Cen. pubescens	[112]
				Chamaecytisus proliferus	[171]
				Cli. ternatea	[112]
				Clitoria sp.	[176]
				Cro. juncea	[112]
				Crotalaria sessiliflora	[112]
				Cytisus scoparius	[112]
				Desmodium intortum	[112]
				Desmodium uncinatum	[112]
				Dumasia truncata	[112]
				Glycine max	[100, 112, 140, 153, 171, 177]
				Glycine soja	[153, 171, 178]
				Indigofera decora	[112]
				Indigofera pseudotinctoria	[112]
				Ind. spicata	[112]
				Kum. striata	[112]
				Lablab purpureus	[112]
				Lens culinaris	[112]

Table 1: cont….

				Les. cuneata	[87, 112]
				Les. cyrtobotrya	[112]
				Les. daurica	[87]
				Les. juncea	[87]
				Lespedeza thunbergii	[112]
				Lespedeza spp.	[162]
				Lupinus angustifolius	[112]
				Lup. luteus	[171]
				Macroptilium atropurpureum	[41, 112, 171]
				Macrotyloma africanum	[176]
				Millettia japonica	[112]
				Mim. pudica	[112]
				Pachyrhizus erosus	[112]
				Pha. acutifolius	[112]
				Phaseolus angularis (Vig. angularis)	[112, 176]
				Pha. vulgaris (ineffective)	[140]
				Psophocarpus tetragonolobus	[112]
				Pue. phaseoloides	[112, 176]
				Rhy. caribaea	[41]
				Rhy. ferulifolia	[41]
				Rhy. minima	[41]
				Sty. hamata	[112]
				Tel. stenopelata	[171]
				Vig. radiata	[102, 112]
				Vig. unguiculata	[24, 102, 112, 176]
				Vig. vexillata	[112]
				Zornia glochidiata	[179]
			Bradyrhizobium jicamae	Pachyrhizus erosus	[180]
			Bradyrhizobium lablabi	Ara. hypogaea	[181]
				Lablab purpureus	[181]
				Vig. unguiculata	[181]
			Bradyrhizobium liaoningense	Acacia saligna	[24]
				Alb. gummifera	[24]
				Erythrina brucei	[24]
				Fai. albida	[24]

Table 1: cont….

					Glycine max	[171, 177, 182]
					Pha. vulgaris	[24]
					Vig. radiata	[102]
					Vig. unguiculata	[102]
					Z. glochidiata	[179]
				Bradyrhizobium lupini	Lup. albus	[183]
				Bradyrhizobium pachyrhizi	Pachyrhizus erosus	[180]
				Bradyrhizobium yuanmingense	Glycine max	[177]
					Glycyrrhiza uralensis	[87]
					Lespedeza cuneata	[162]
					Med. sativa	[87]
					Mel. albus	[87]
					Vig. radiata	[102]
					Vig. unguiculata	[24, 87, 102]
					Z. glochidiata	[179]
				Bradyrhizobium spp.	Acacia albida	[29]
					Acacia longifolia	[184]
					Acacia mangium	[29]
					Acacia saligna	[24]
					Aeschynomene afraspera	[138]
					Aeschynomene fluminensis	[138]
					Aes. indica	[112]
					Alb. gummifera	[24]
					Aly. vaginalis	[112]
					Cas. nomame	[112]
					Centrosema pascuorum	[112]
					Cen. pubescens	[112]
					Clitoria sp.	[141]
					Cro. juncea	[138]
					Cytisus balansae	[185]
					Cytisus multiflorus	[185]
					Cyt. scoparius	[185]
					Cytisus striatus	[185]
					Dalbergia sp.	[27]
					Der. elliptica	[42, 44]
					Derris trifoliata	[112]

Table 1: cont....

				Ery. brucei	[24]
				Genista hystrix	[185]
				Glycine max	[138]
				Ind. arrecta	[138]
				Indigofera brevicalyx	[138]
				Indigofera hirsuta	[138]
				Ind. spicata	[112]
				Indigofera suffruticosa	[138]
				Indigofera sp.	[138]
				Laburnum anagyroides	[141]
				Lupinus luteus	[99, 100]
				Lupinus spp.	[100, 186]
				Macrotyloma africanum	[99, 138]
				Macrotyloma sp.	[138]
				Mil. ferruginea	[24]
				Neo. wightii	[112]
				Ornithopus compressus	[186]
				Paraserianthes falcataria	[29]
				Parasponia andersonii	[114]
				Parasponia sp.	[170]
				Pue. mirifica	[44, 69]
				Pue. phaseoloides	[112]
				Retama sphaerocarpa	[185]
				Sam. saman	[112]
				Sty. guianensis	[112]
				Sty. hamata	[112]
				Vig. radiata	[138, 187]
				Vig. unguiculata	[138, 141]
				Z. glochidiata	[179]
	Phyllobacte-riaceae	Mesorhizobium	Mesorhizobium albiziae	Albizia kalkora	[188]
				Lot. corniculatus	[189]
				Lot. tenuis	[189]
			Mesorhizobium alhagi	Alhagi sparsifolia	[190]
				Lot. corniculatus	[189]
				Lot. tenuis	[189]
			Mesorhizobium amorphae	Amo. fruticosa	[83]
				Lespedeza spp.	[162]

Table 1: cont....

			Mesorhizobium australicum	*Ast. membranaceus*	[191]
				Biserrula pelecinus	[191]
				Macroptilium atropurpureum	[191]
			Mesorhizobium camelthorni	*Alh. sparsifolia*	[192]
			Mesorhizobium caraganae	*Ast. adsurgens*	[193]
				Car. intermedia	[193]
				Car. microphylla	[193]
				Glycyrrhiza uralensis	[193]
				Pha. vulgaris	[193]
			Mesorhizobium chacoense	*Acacia abyssinica*	[24]
				Lot. corniculatus	[189]
				Lot. tenuis	[189]
				Prosopis alba	[194]
			Mesorhizobium ciceri (formerly *Rhizobium ciceri*) [16]	*Bis. pelecinus*	[195]
				Cic. arietinum	[100, 195, 196]
			Mesorhizobium gobiense	*Astragalus filicaulis*	[197]
				Glycyrrhiza uralensis	[197]
				Lot. corniculatus	[197]
				Lotus frondosus	[197]
				Lot. tenuis	[197]
				Oxytropis glabra	[197]
				Rob. pseudoacacia	[197]
			Mesorhizobium huakuii (formerly *Rhizobium huakuii*) [16]	*Astragalus sinicus*	[198, 199]
				Lespedeza spp.	[162]
			Mesorhizobium loti (formerly *Rhizobium loti*) [16]	*Acacia senegal*	[200]
				Acacia seyal	[200]
				Amo. fruticosa	[112]
				Ast. adsurgens	[112]
				Ast. membranaceus	[112]
				Ast. sinicus	[112]
				Cic. arietinum	[112]
				Leu. leucocephala	[112]
				Lot. corniculatus	[100, 112, 200, 201]
				Lotus divaricatus	[100, 116, 161, 201]

Table 1: cont….

					Lot. japonicus	[202]
					Lot. tenuis	[100, 201]
					Mim. pudica	[112]
					Neptunia oleracea	[200]
					Rob. pseudoacacia	[52, 112]
					Ses. glandiflora	[112]
					Sophora flavescens	[112]
					Thermopsis lupinoides (*Thermopsis fabacea*)	[112]
					Trifolium lupinaster	[112]
					Wisteria floribunda	[112]
				Mesorhizobium mediterraneum (formerly *Rhizobium mediterraneum*) [16]	*Cic. arietinum*	[76, 100]
					Glycyrrhiza glabra (ineffective/effective)	[94]
					Glycyrrhiza uralensis	[94]
				Mesorhizobium metallidurans	*Ant. vulneraria*	[203]
				Mesorhizobium opportunistum	*Ast. membranaceus*	[191]
					Ast. adsurgens	[191]
					Bis. pelecinus (ineffective)	[191]
					Lotus peregrinus	[191]
					Macroptilium atropurpureum	[191]
				Mesorhizobium plurifarium	*Acacia abyssinica*	[24]
					Acacia nilotica	[200]
					Acacia senegal	[24, 200]
					Acacia seyal	[200]
					Acacia tortilis	[24, 200]
					Leu. leucocephala	[200, 204]
					Nep. oleracea	[200]
					Ses. herbacea	[204]
					Ses. sesban	[24]
				Mesorhizobium robiniae	*Rob. pseudoacacia*	[205]
				Mesorhizobium septentrionale	*Ast. adsurgens*	[206]
					Glycine max	[206]
					Leu. leucocephala	[206]
					Lot. corniculatus	[206]
					Macroptilium atropurpureum	[206]

					Pha. vulgaris	[206]
				Mesorhizobium shangrilense	*Ast. adsurgens*	[207]
					Car. intermedia	[207]
					Car. microphylla	[207]
					Glycyrrhiza uralensis	[207]
					Pha. vulgaris	[207]
					Vig. radiata	[207]
					Vig. unguiculata	[207]
				Mesorhizobium tarimense	*Glycyrrhiza uralensis*	[197]
					Lot. corniculatus	[189, 197]
					Lot. frondosus	[197]
					Lot. tenuis	[189, 197]
					Oxytropis glabra	[197]
					Rob. pseudoacacia	[197]
				Mesorhizobium temperatum	*Ast. adsurgens*	[206]
					Glycine max	[206]
					Leu. leucocephala	[206]
					Lot. corniculatus	[206]
					Med. sativa	[206]
					Pha. vulgaris	[206]
					Vig. unguiculata	[206]
				Mesorhizobium tianshanense (formerly *Rhizobium tianshanense*) [16]	*Caragana polourensis*	[208, 209]
					Cic. arietimum	[210]
					Glycine max	[208]
					Glycyrrhiza glabra	[94]
					Glycyrrhiza pallidiflora	[208, 209]
					Glycyrrhiza uralensis	[94, 208, 209]
					Hal. holodendron	[208, 209]
					Lot. corniculatus	[189, 208]
					Lot. tenuis	[189, 208]
					Sophora alopecuroides	[208, 209]
					Swainsonia salsula	[208]
				Mesorhizobium spp.	*Acacia tortilis*	[24]
					Astragalus spp.	[100]
					Cic. arietinum	[76, 100]
					Coronilla varia	[99, 100]
					Dalbergia sp.	[27]
					Glycyrrhiza glabra	[94]

Table 1: cont….

				Glycyrrhiza uralensis	[94]
				Lot. tenuis	[143]
				Onobrychis spp.	[100]
				Oxytropis spp.	[100]
				Ses. sesban	[24]
		Phyllobacterium	*Phyllobacterium ifriqiyense*	*Astragalus algerianus*	[28]
				Lathyrus numidicus	[28]
			Phyllobacterium leguminum	*Arg. uniflorum*	[28]
				Ast. algerianus	[28]
			Phyllobacterium trifolii	*Lup. albus*	[26]
				Trifolium pratense	[26]
				Trifolium repens	[26]
			Phyllobacterium spp.	*Dalbergia* spp.	[27]
	Hyphomicro-biaceae	*Azorhizobium*	*Azorhizobium caulinodans*	*Leu. leucocephala*	[40]
				Pha. vulgaris (ineffective)	[40]
				Ses. rostrata	[13, 211]
				Vig. unguiculata (ineffective)	[40]
			Azorhizobium doebereinerae	*Leu. leucocephala*	[40]
				Macroptilium atropurpureum (ineffective)	[19, 40]
				Pha. vulgaris (ineffective)	[19, 40]
				Ses. rostrata (stem pseudonodules/inffective root nodules)	[19]
				Sesbania virgata	[19, 40]
				Vig. unguiculata (ineffective)	[40]
			Azorhizobium spp.	*Dalbergia* sp.	[27]
				Ses. rostrata	[138]
				Z. glochidiata (ineffective)	[179]
		Devosia	*Devosia neptuniae*	*Nep. natans*	[25]
	Methylobac-teriaceae	*Methylobacterium*	*Methylobacterium nodulans*	*Crotalaria glaucoides*	[23]
				Crotalaria perrottetii	[23]
				Crotalaria podocarpa	[23]
			Methylobacterium spp.	*Fai. albida*	[24]
				Pha. vulgaris	[24]
				Ses. sesban	[24]
				Vig. unguiculata	[24]

Table 1: cont….

		Microvirga	*Microvirga lotononidis*	*Listia angolensis*	[34]
			Microvirga lupini	*Lupinus texensis*	[34]
			Microvirga zambiensis	*Lis. angolensis*	[34]
	Brucellaceae	*Ochrobactrum*	*Ochrobactrum cytisi*	*Cyt. scoparius*	[31]
			Ochrobactrum lupini	*Lup. albus*	[30]
				Lupinus honoratus	[30]
			Ochrobactrum sp.	*Acacia albida*	[29]
				Acacia mangium	[29]
				Paraserianthes falcataria	[29]
Beta-Proteo bacteria	Burkholderia-ceae	*Cupriavidus/ Ralstonia*	*Cupriavidus taiwanensis* (formerly *Ralstonia taiwanensis*) [212]	*Macroptilium atropupureum*	[41]
				Mim. diplotricha	[36, 37]
				Mim. pigra	[39]
				Mim. pudica	[36, 37]
				Rhy. caribaea	[41]
				Rhy. ferulifolia	[41]
				Rhy. minima	[41]
			Cupriavidus spp. and *Ralstonia* spp.	*Dalbergia* sp.	[27]
				Ind. tinctoria	[42, 44]
				Leu. leucocephala	[40]
				Macroptilium atropurpureum	[40]
				Mim. diplotricha	[37]
				Mim. pigra	[35]
				Mim. pudica	[35, 37]
				Mimosa spp.	[43]
				Pha. vulgaris	[40]
		Burkholderia	*Burkholderia caribensis*	*Mim. diplotricha*	[38]
				Mim. pudica	[37]
			Burkholderia cepacia	*Dalbergia* sp.	[27]
			Burkholderia mimosarum	*Mim. pigra*	[46]
			Burkholderia nodosa	*Mimosa bimucronata*	[47]
				Mimosa scabrella	[47]
			Burkholderia phymatum	*Acacia bilimeki*	[53]
				Acacia cochliacantha (*Vachellia campechiana*)	[53]

Table 1: cont….

				Leucaena glauca (*Leucaena leucocephala*)	[53]
				Macherium lunatum	[51, 54]
				Mim. pigra	[53]
				Pha. vulgaris	[53]
				Prosopis laevigata	[53]
				Rhy. caribaea	[41]
				Rhy. ferulifolia	[41]
				Rhy. minima	[41]
			Burkholderia sabiae	*Mimosa caesalpiniifolia*	[48]
			Burkholderia tuberum	*Aspalathus callosa* (original host but no nodulation)	[49]
				Aspalathus carnosa	[49, 51, 54]
				Cyclopia falcata	[49]
				Cyclopia galiodes	[49]
				Cyclopia genistoides	[49]
				Cyclopia intermedia	[49]
				Cyclopia pubescens	[49]
				Macroptilium atropurpureum	[49]
			Burkholderia spp.	*Abarema macradenia*	[213]
				Dalbergia odorifera	[50]
				Dalbergia spp.	[27]
				Macroptilium atropupureum	[41, 213]
				Mimosa acutistipula	[45]
				Mim. bimucronata	[45]
				Mimosa caesalpiniaefolia (*Mimosa caesalpiniifolia*)	[45]
				Mimosa camporum	[45]
				Mimosa casta	[213]
				Mimosa flocculosa	[45]
				Mimosa laticifera	[45]
				Mim. pigra	[45, 213]
				Mim. pudica	[35, 213]
				Mim. scabrella	[45]
				Mimosa tenuiflora	[45]
				Mimosa spp.	[43]
				Rhy. ferulifolia	[41]
				Rhy. minima	[41]

Table 1: cont….

				Rob. pseudoacacia	[52]
	Oxalobacte-riaceae	*Herbaspirillum*	*Herbaspirillum lusitanum*	*Pha. vulgaris*	[55]
Gamma-Proteo bacteria	Pseudomona-daceae	*Pseudomonas*	*Pseudomonas* sp.	*Rob. pseudoacacia*	[52]

Pseudonodules: irregular swellings on root surface but not identifiable as true nodules; ineffective: nodules were formed but no nitrogen fixed; effective: nitrogen-fixing nodules were formed.
Not all strains belonging to the same species form nodules with the same hosts.

REFERENCES

[1]   Novikova NI, Pavlova EA, Vorobjev NI, Limeshchenko EV. Numerical taxonomy of *Rhizobium* strains from legumes of the temperate zone. Int J Syst Bacteriol 1994; 44: 734-42.

[2]   Hartmann A, Amarger N. Genotypic diversity of an indigenous *Rhizobium meliloti* field population assessed by plasmid profiles, DNA fingerprinting and insertion sequence typing. Can J Microbiol 1991; 37: 600-08.

[3]   Moris M, Braeken K, Schoeters E, *et al.* Effective symbiosis between *Rhizobium etli* and *Phaseolus vulgaris* requires the alarmone ppGpp. J Bacteriol 2005; 187: 5460-69.

[4]   Zahran HH. *Rhizobium*-legume symbiosis and nitrogen fixation under severe conditions in an arid climate. Microbiol Mol Biol Rev 1999; 63: 968-89.

[5]   de Philip P, Boistard P, Schluter A, *et al.* Developmental and metabolic regulation of nitrogen fixation gene expression in *Rhizobium meliloti*. Can J Microbiol 1992; 38: 467-74.

[6]   Doyle JJ. Phylogeny of the legume family: an approach to understanding the origins of nodulation. Annu Rev Ecol Syst 1994; 25: 325-49.

[7]   Gentry AH. Changes in plant community diversity and floristic composition on environmental and geographical gradients. Ann Mo Bot Gard 1988; 75: 1-34.

[8]   Jordan DC. Family III. Rhizobiaceae Conn 1938, 321[AL]. In: Krieg NR, Holt JG, Eds. Bergey's Manual of Systematic Bacteriology, Vol. 1. Baltimore, The Williams and Wilkins Co., 1984; pp. 235-44.

[9]   Somasegaran P, Hoben HJ. Handbook for Rhizobia: Methods in Legume-*Rhizobium* Technology. NIFTAL Project, University of Hawaii, Paia, USA. 1994.

[10]   Chen WX, Yan GH, Li JL. Numerical taxonomic study of fast-growing soybean rhizobia and a proposal that *Rhizobium fredii* be assigned to *Sinorhizobium* gen. nov. Int J Syst Bacteriol 1988; 38: 392-97.

[11]   Terefework Z, Nick G, Suomalainen S, *et al.* Phylogeny of *Rhizobium galegae* with respect to other rhizobia and agrobacteria. Int J Syst Bacteriol 1998; 48: 349-56.

[12]   The Judicial Commission of the International Committee on Systematics of Prokaryotes. The genus name *Sinorhizobium* Chen *et al.* 1988 is a later synonym of *Ensifer* Casida 1982 and is not conserved over the latter genus name and the species name '*Sinorhizobium adhaerens*' is not validly published. Opinion 84. Int J Syst Evol Microbiol 2008; 58: 1973.

[13]   Dreyfus B, Garcia JL, Gillis M. Characterization of *Azorhizobium caulinodans* gen. nov., sp. nov., a stem-nodulating nitrogen-fixing bacterium isolated from *Sesbania rostrata*. Int J Syst Bacteriol 1988; 38: 89-98.

[14]   de Lajudie P, Laurent-Fulele E, Willems A, *et al. Allorhizobium undicola* gen. nov., sp. nov., nitrogen-fixing bacteria that efficiently nodulate *Neptunia natans* in Senegal. Int J Syst Bacteriol 1998; 48: 1277-90.

[15]   Young JM, Kuykendall LD, Martinez-Romero E, *et al.* A revision of *Rhizobium* Frank 1889, with an emended description of the genus and the inclusion of all species of *Agrobacterium* Conn 1942 and *Allorhizobium undicola* de Lajudie *et al.* 1998 as new combinations: *Rhizobium radiobacter, R. rhizogenes, R. rubi, R. undicola* and *R. vitis.* Int J Syst Evol Microbiol 2001; 51: 89-03.

[16]   Jarvis BDW, van Berkum P, Chen WX, *et al.* Transfer of *Rhizobium loti, Rhizobium huakuii, Rhizobium ciceri, Rhizobium mediterraneum* and *Rhizobium tianshanense* to *Mesorhizobium* gen. nov. Int J Syst Bacteriol 1997; 47: 895-98.

[17]   van Berkum P, Eardly BD. The aquatic budding bacterium *Blastobacter denitrificans* is a nitrogen-fixing symbiont of *Aeschynomene indica.* Appl Envir Microbiol 2002; 68: 1132-36.

[18]   van Berkum P, Leibold JM, Eardly BD. Proposal for combining *Bradyrhizobium* spp. (*Aeschynomene indica*) with *Blastobacter denitrificans* and to transfer *Blastobacter denitrificans* (Hirsh and Muller, 1985) to the genus *Bradyrhizobium* as *Bradyrhizobium denitrificans* (comb. nov.). Syst Appl Microbiol 2006; 29: 207-15.

[19]   Moreira FMS, Cruz L, de Faria SM, *et al. Azorhizobium doebereinerae* sp. nov. microsymbiont of *Sesbania virgata* (Caz.) Pers. Syst Appl Microbiol 2006; 29: 197-06.

[20]   Boivin C, Ndoye I, Lortet G, *et al.* The *Sesbania* root symbionts *Sinorhizobium saheli* and *S. teranga* bv. sesbaniae can form stem nodules on *Sesbania rostrata*, although they are less adapted to stem nodulation than *Azorhizobium caulinodans.* Appl Envir Microbiol 1997; 63: 1040-47.

[21]   Rinaudo G, Orenga S, Fernandez MP, *et al.* DNA homologies among members of the genus *Azorhizobium* and other stem- and root-nodulating bacteria isolated from the tropical legume *Sesbania rostrata.* Int J Syst Evol Microbiol 1991; 41: 114-20.

[22]   Giraud E, Hannibal L, Fardoux J, *et al.* Effect of *Bradyrhizobium* photosynthesis on stem nodulation of *Aeschynomene sensitiva.* Proc Natl Acad Sci USA. 2000; 97: 14795-00.

[23]   Sy A, Giraud E, Jourand P, *et al.* Methylotrophic *Methylobacterium* bacteria nodulate and fix nitrogen in symbiosis with legumes. J Bacteriol 2001; 183: 214-20.

[24]   Wolde-meskel E, Terefework Z, Frostegard A, Lindstrom K. Genetic diversity and phylogeny of rhizobia isolated from agroforestry legume species in southern Ethiopia. Int J Syst Evol Microbiol 2005; 55: 1439-52.

[25]   Rivas R, Valazquez E, Willems A, *et al.* A new species of *Devosia* that forms a unique nitrogen-fixing root-nodule symbiosis with the aquatic legume *Neptunia natans* (L.f.) Druce. Appl Envir Microbiol 2002; 68: 5217-22.

[26]   Valverde A, Velazquez E, Fernandez-Santos F, *et al. Phyllobacterium trifolii* sp. nov., nodulating *Trifolium* and *Lupinus* in Spanish soils. Int J Syst Evol Microbiol 2005; 55: 1985-89.

[27]   Rasolomampianina R, Bailly X, Fetiarison R, *et al.* Nitrogen-fixing nodules from rose wood legume trees (*Dalbergia* spp.) endemic to Madagascar host seven different genera belonging to alpha- and beta-Proteobacteria. Mol Ecol 2005; 14: 4135-46.

[28]   Mantelin S, Fischer-Le Saux M, Zakhia F, *et al.* Emended description of the genus *Phyllobacterium* and description of four novel species associated with plant roots: *Phyllobacterium bourgognense* sp. nov., *Phyllobacterium ifriqiyense* sp. nov., *Phyllobacterium leguminum* sp. nov. and *Phyllobacterium brassicacearum* sp. nov. Int J Syst Evol Microbiol 2006; 56: 827-39.

[29]   Ngom A, Nakagawa Y, Sawada H, *et al.* A novel symbiotic nitrogen-fixing member of the *Ochrobactrum* clade isolated from root nodules of *Acacia mangium.* J Gen Appl Microbiol 2004; 50: 17-27.

[30]   Trujillo ME, Willems A, Abril A, *et al.* Nodulation of *Lupinus albus* by strains of *Ochrobactrum lupini* sp. nov. Appl Envir Microbiol 2005; 71: 1318-27.

[31]   Zurdo-Pineiro JL, Rivas R, Trujillo ME, *et al. Ochrobactrum cytisi* sp. nov., isolated from nodules of *Cytisus scoparius* in Spain. Int J Syst Evol Microbiol 2007; 57: 784-88.

[32]   Andam CP, Parker MA. Novel alphaproteobacterial root nodule symbiont associated with *Lupinus texensis*. Appl Envir Microbiol 2007; 73: 5687-91.

[33]   Lin DX, Wang ET, Tang H, *et al. Shinella kummerowiae* sp. nov., a symbiotic bacterium isolated from root nodules of the herbal legume *Kummerowia stipulacea*. Int J Syst Evol Microbiol 2008; 58: 1409-13.

[34]   Ardley JK, Parker MA, de Meyer SE, *et al. Microvirga lupini* sp. nov., *Microvirga lotononidis* sp. nov. and *Microvirga zambiensis* sp. nov. are alphaproteobacterial root nodule bacteria that specifically nodulate and fix nitrogen with geographically and taxonomically separate legume hosts. Int J Syst Evol Microbiol 2011; doi: 10.1099/ijs.0.035097-0.

[35]   Barrett CF, Parker MA. Coexistence of *Burkholderia*, *Cupriavidus* and *Rhizobium* sp. nodule bacteria on two *Mimosa* spp. in Costa Rica. Appl Envir Microbiol 2006; 72: 1198-06.

[36]   Chen WM, Laevens S, Lee TM, *et al. Ralstonia taiwanensis* sp. nov., isolated from root nodules of *Mimosa* species and sputum of a cystic fibrosis patient. Int J Syst Evol Microbiol 2001; 51: 1729-35.

[37]   Chen WM, James EK, Prescott AR, *et al.* Nodulation of *Mimosa* spp. by the β-proteobacterium *Ralstonia taiwanensis.* Mol Plant-Microbe Interact 2003; 16: 1051-61.

[38]   Chen WM, Moulin L, Bontemps C, *et al.* Legume symbiotic nitrogen fixation by β-Proteobacteria is widespread in nature. J Bacteriol 2003; 185: 7266-72.

[39]   Chen WM, James EK, Chou JH, *et al.* β-rhizobia from *Mimosa pigra*, a newly discovered invasive plant in Taiwan. New Phytologist 2005; 168: 671-75.

[40]   Florentino LA, Guimaraes AP, Rufini M, *et al. Sesbania virgata* stimulates the occurrence of its microsymbiont in soils but does not inhibit microsymbionts of other species. Sci Agric (Piracicaba, Braz) 2009; 66: 667-76.

[41]   Garau G, Yates RJ, Deiana P, Howieson JG. Novel strains of nodulating *Burkholderia* have a role in nitrogen fixation with papilionoid herbaceous legumes adapted to acid, infertile soils. Soil Biol Biochem 2009; 41: 125-34.

[42]   Leelahawonge C, Nuntagij A, Teaumroong N, *et al.* Characterization of root-nodule bacteria isolated from the medicinal legume *Indigofera tinctoria*. Ann Microbiol 2010; 60: 65-74.

[43]   Liu XY, Wu W, Wang ET, *et al.* Phylogenetic relationships and diversity of β-rhizobia associated with *Mimosa* species grown in Sishuangbanna, China. Int J Syst Evol Microbiol 2011; 61: 334-42.

[44]   Pongsilp N, Nuntagij A. Genetic diversity and metabolites production of root-nodule bacteria isolated from medicinal legumes *Indigofera tinctoria*, *Pueraria mirifica* and *Derris elliptica* Benth. grown in different geographic origins across Thailand. Amer-Eur J Agric Envir Sci 2009; 6: 26-34.

[45]   Chen WM, de Faria SM, Straliotto R, *et al.* Proof that *Burkholderia* strains form effective symbioses with legumes: a study of novel *Mimosa*-nodulating strains from South America. Appl Envir Microbiol 2005; 71: 7461-71.

[46]   Chen WM, James EK, Coenye T, *et al. Burkholderia mimosarum* sp. nov., isolated from root nodules of *Mimosa* spp. from Taiwan and South America. Int J Syst Evol Microbiol 2006; 56: 1847-51.

[47]   Chen WM, de Faria SM, James EK, *et al. Burkholderia nodosa* sp. nov., isolated from root nodules of the woody Brazilian legumes *Mimosa bimucronata* and *Mimosa scabrella*. Int J Syst Evol Microbiol 2007; 57: 1055-69.

[48]   Chen WM, de Faria SM, Chou JH, *et al. Burkholderia sabiae* sp. nov., isolated from root nodules of *Mimosa caesalpiniifolia.* Int J Syst Evol Microbiol 2008; 58: 2174-79.

[49]   Elliott GN, Chen WM, Bontemps C, *et al.* Nodulation of *Cyclopia* spp. (Leguminosae, Papilionoideae) by *Burkholderia tuberum,* Ann Bot 2007; 100: 1403-11.

[50]   Lu JK, He XH, Huang LB, *et al.* Two *Burkholderia* strains from nodules of *Dalbergia odorifera* T. Chen in Hainan Island, southern China. New Forests 2011; doi: 10.1007/s11056-011-9290-8.

[51]   Moulin L, Munive A, Dreyfus B, Boivin-Masson C. Nodulation of legumes by members of the β-subclass of proteobacteria. Nature 2001; 411: 948-50.

[52]   Shiraishi A, Matsushita N, Hougetsu T. Nodulation in black locust by the Gammaproteobacteria *Pseudomonas* sp. and the Betaproteobacteria *Burkholderia* sp. Syst Appl Microbiol 2010; 33: 269-74.

[53]   Talbi C, Delgado MJ, Girard L, *et al. Burkholderia phymatum* strains capable of nodulating *Phaseolus vulgaris* are present in Moroccan soils. Appl Envir Microbiol 2010; 76: 4587-91.

[54]   Vandamme P, Goris J, Chen WM, *et al. Burkholderia tuberum* sp. nov. and *Burkholderia phymatum* sp. nov. nodulate the roots of tropical legumes. Syst Appl Microbiol 2002; 25: 507-12.

[55]   Valverde A, Velazquez E, Gutierrez C, *et al. Herbaspirillum lusitanum* sp. nov., a novel nitrogen-fixing bacterium associated with root nodules of *Phaseolus vulgaris.* Int J Syst Evol Microbiol 2003; 53: 1979-83.

[56]   Kaneko T, Nakamura Y, Sato S, *et al.* Complete genome structure of the nitrogen-fixing symbiotic bacterium *Mesorhizobium loti.* DNA Res 2000; 7: 331-38.

[57]   Sullivan JT, Ronson CW. Evolution of rhizobia by acquisition of a 500-kb symbiosis island that integrates into a phe-tRNA gene. Proc Natl Acad Sci USA. 1998; 95: 5145-49.

[58]   Laguerre G, Geniaux E, Mazurier SI, *et al.* Conformity and diversity among field isolates of *Rhizobium leguminusarum* bv. *viciae,* bv. *trifolii* and bv. *phaseoli* revealed by DNA hybridization using chromosome and plasmid probes. Can J Microbiol 1992; 39: 412-19.

[59]   Louvrier P, Laguerre G, Amarger N. Distribution of symbiotic genotypes in *Rhizobium leguminosarum* biovar viciae populations isolated directly from soils. Appl Envir Microbiol 1996; 62: 4202-05.

[60]   Schofield PR, Gibson AH, Dudman WF, Watson JM. Evidence for genetic exchange and recombination of *Rhizobium* symbiotic plasmid in a soil population. Appl Envir Microbiol 1987; 53: 2942-47.

[61]   Young JPW, Wexler M. Sym plasmid and chromosomal genotypes are correlated in field populations of *Rhizobium leguminosarum.* J Gen Microbiol 1988; 134: 2731-39.

[62]   Marchetti M, Capela D, Glew M, *et al.* Experimental evolution of a plant pathogen into a legume symbiont. PLoS Biol 2010; 8: e1000280

[63]   Rogel MA, Hernandez-Lucas I, Kuykendall LD, *et al.* Nitrogen-fixing nodules with *Ensifer adhaerens* harboring *Rhizobium tropici* symbiotic plasmids. Appl Envir Microbiol 2001; 67: 3264-68.

[64]   Pongsilp N, Teaumroong N, Nuntagij A, *et al.* Genetic structure of indigenous non-nodulating and nodulating populations of *Bradyrhizobium* in soils from Thailand. Symbiosis 2002; 33: 39-58.

[65]   Segovia L, Pinero D, Palacios R, Martinez-Romero E. Genetic structure of a soil population of nonsymbiotic *Rhizobium leguminosarum.* Appl Envir Microbiol 1991; 57: 426-33.

[66]   Laguerre G, Bardin M, Amarger N. Isolation from soil of symbiotic and nonsymbiotic *Rhizobium leguminosarum* by DNA hybridization. Can J Microbiol 1993; 39: 1142-49.

[67]   Soberon-Chavez G, Najera R. Isolation from soil of *Rhizobium leguminosarum* lacking symbiotic information. Can J Microbiol 1989; 35: 464-68.

[68]   Sullivan JT, Eardly BD, van Berkum P, Ronson CW. Four unnamed species of nonsymbiotic rhizobia isolated from the rhizosphere of *Lotus corniculatus*. Appl Envir Microbiol 1996; 62: 2818-25.

[69]   Pongsilp N, Leelahawonge C, Nuntagij A, *et al.* Characterization of *Pueraria mirifica*-nodulating rhizobia present in Thai soil. Afr J Microbiol Res 2010; 4: 1307-13.

[70]   Bais HP, Weir TL, Perry LG, *et al.* The role of root exudates in rhizosphere interactions with plants and other organisms. Annu Rev Plant Biol 2006; 57: 233-66.

[71]   Weisskopf L, Abou-Mansour E, Fromin N, *et al.* White lupin has developed a complex strategy to limit microbial degradation of secreted citrate required for phosphate acquisition. Plant Cell Envir 2006; 29: 919-27.

[72]   Fuentes JB, Abe M, Uchiumi T, *et al.* Symbiotic root nodule bactetia isolated from yam bean (*Pachyrhizus erosus*). J Gen Appl Microbiol 2002; 48: 181-91.

[73]   Pueppke SG, Broughton WJ. *Rhizobium* sp. strain NGR234 and *R. fredii* USDA 257 share exceptionally broad, nested host ranges. Mol Plant-Microbe Interact 1999; 12: 293-18.

[74]   Niemann S, Puhler A, Tichy HV, *et al.* Evaluation of the resolving power of three different DNA fingerprinting methods to discriminate among isolates of a natural *Rhizobium meliloti* population. J Appl Microbiol 1997; 82: 477-84.

[75]   Demezas DH, Reardon TB, Watson JM, Gibson AH. Genetic diversity among *Rhizobium leguminosarum* bv. *trifolii* strains revealed by allozyme and restriction fragment length polymorphism analyses. Appl Envir Microbiol 1991; 57: 3489-95.

[76]   Nour SM, Cleyet-Marel JC, Normand P, Fernandez MP. Genomic heterogeneity of strains nodulating chickpeas (*Cicer arietinum* L.) and description of *Rhizobium mediterraneum* sp. nov. Int J Syst Bacteriol 1995; 45: 640-48.

[77]   Bromfield ESP, Behara AMP, Singh RS, Barran LR. Genetic variation in local populations of *Sinorhizobium meliloti*. Soil Biol Biochem 1998; 30: 1707-16.

[78]   Aguilar OM, Lopez MV, Riccillo PM, *et al.* Prevalence of the *Rhizobium etli*-like allele in genes coding for 16S rRNA among the indigenous rhizobial populations found associated with wild beans from the southern Andes in Argentina. Appl Envir Microbiol 1998; 64: 3520-24.

[79]   Lafay B, Burdon JJ. Molecular diversity of rhizobia occurring on native shrubby legumes in southeastern Australia. Appl Envir Microbiol 1998; 64: 3989-97.

[80]   Khbaya B, Neyra M, Normand P, *et al.* Genetic diversity and phylogeny of rhizobia that nodulate *Acacia* spp. in Morocco assessed by analysis of rRNA genes. Appl Envir Microbiol 1998; 64: 4912-17.

[81]   van Berkum P, Beyene D, Bao G, *et al. Rhizobium mongolense* sp. nov. is one of three rhizobial genotypes identified which nodulate and form nitrogen-fixing symbioses with *Medicago ruthenica* [(L.) Ledebour]. Int J Syst Evol Microbiol 1998; 48: 13-22.

[82]   Wang ET, Rogel MA, Garcia-de los Santos A, *et al. Rhizobium etli* bv. mimosae, a novel biovar isolated from *Mimosa affinis*. Int J Syst Bacteriol 1999; 49: 1479-91.

[83]   Wang ET, Rogel MA, Sui XH, *et al. Mesorhizobium amorphae*, a rhizobial species that nodulates *Amorpha fruticosa*, is native to American soils. Arch Microbiol 1999; 178: 301-05.

[84]   Segundo E, Martinez-Abarca F, van Dillewijn P, *et al.* Characterisation of symbiotically efficient alfalfa-nodulating rhizobia isolated from acid soils of Argentina and Uruguay. FEMS Microbiol Ecol 1999; 28: 169-76.

[85]   Carelli M, Gnocchi S, Fancelli S, *et al.* Genetic diversity and dynamics of *Sinorhizobium meliloti* populations nodulating different alfalfa cultivars in Italian soils. Appl Envir Microbiol 2000; 66: 4785-89.

[86]   Diouf A, de Lajudie P, Neyra M, *et al.* Polyphasic characterization of rhizobia that nodulate *Phaseolus vulgaris* in West Africa (Senegal and Gambia). Int J Syst Evol Microbiol 2000; 50: 159-70.

[87]   Yao ZY, Kan FL, Wang ET, *et al.* Characterization of rhizobia that nodulate legume species of the genus *Lespedeza* and description of *Bradyrhizobium yuanmingense* sp. nov. Int J Syst Evol Microbiol 2002; 52: 2219-30.

[88]   Wei GH, Wang ET, Tan ZY, *et al. Rhizobium indigoferae* sp. nov. and *Sinorhizobium kummerowiae* sp. nov., respectively isolated from *Indigofera* spp. and *Kummerowia stipulacea.* Int J Syst Evol Microbiol 2002; 52: 2231-39.

[89]   Wei GH, Tan ZY, Zhu ME, *et al.* Characterization of rhizobia isolated from legume species within the genera *Astragalus* and *Lespedeza* grown in the Loess Plateau region of China and description of *Rhizobium loessense* sp. nov. Int J Syst Evol Microbiol 2003; 53: 1575-83.

[90]   Andronov EE, Terefework Z, Roumiantseva ML, *et al.* Symbiotic and genetic diversity of *Rhizobium galegae* isolates collected from the *Galega orientalis* gene center in the Caucasus. Appl Envir Microbiol 2003; 69: 1067-74.

[91]   Silva C, Vinuesa P, Eguiarte LE, *et al. Rhizobium etli* and *Rhizobium gallicum* nodulate common bean (*Phaseolus vulgaris*) in a traditionally managed Milpa Plot in Mexico: population genetics and biogeographic implications. Appl Envir Microbiol 2003; 69: 884-93.

[92]   Bromfield ESP, Tambong JT, Cloutier S, *et al. Ensifer, Phyllobacterium* and *Rhizobium* species occupy nodules of *Medicago sativa* (alfalfa) and *Melilotus alba* (sweet clover) grown at a Canadian site without a history of cultivation. Microbiol 2010; 156: 505-20.

[93]   Lu YL, Chen WF, Han LL, *et al. Rhizobium alkalisoli* sp. nov., isolated from Caragana intermedia growing in saline-alkaline soils in the north of China. Int J Syst Evol Microbiol 2009; 59: 3006-11

[94]   Li L, Sinkko H, Montonen L, *et al.* Biogeography of symbiotic and other endophytic bacteria isolated from medicinal *Glycyrrhiza* species in China. FEMS Microbiol Ecol 2012; 79: 46-68.

[95]   Garcia-Fraile P, Rivas R, Willems A, *et al. Rhizobium cellulosilyticum* sp. nov. isolated from sawdust of *Populus alba.* Int J Syst Evol Microbiol 2007; 57: 844-48.

[96]   Quan ZX, Bae HS, Baek JH, *et al. Rhizobium daejeonense* sp. nov., isolated from a cyanide treatment bioreactor. Int J Syst Evol Microbiol 2005; 55: 2543-49.

[97]   Tian CF, Wang ET, Wu LJ, *et al. Rhizobium fabae* sp. nov., a bacterium that nodulates *Vicia faba.* Int J Syst Evol Microbiol 2008; 58: 2871-75.

[98]   Segovia L, Young JPW, Martinez-Romero E. Reclassification of American *Rhizobium leguminosarum* biovar *phaseoli* type I strains as *Rhizobium etli* sp. nov. Int J Syst Bacteriol 1993; 43: 374-77.

[99]   Laguerre G, Allard MR, Revoy F, Amarger N. Rapid identification of rhizobia by restriction fragment length polymorphism analysis of PCR-amplified 16S rRNA genes. Appl Envir Microbiol 1994; 60: 56-63.

[100]   Laguerre G, van Berkum P, Amarger N, Prevost D. Genetic diversity of rhizobial symbionts isolated from legume species within the genera *Astragalus, Oxytropis* and *Onobrychis.* Appl Envir Microbiol 1997; 63: 4748-58.

[101]   Shamseldin A, El-Saadani M, Sadowsky MJ, An CS. Rapid identification and discrimination among Egyptian genotypes of *Rhizobium leguminosarum* bv. *viciae* and *Sinorhizobium meliloti* nodulating faba bean (*Vicia faba* L.) by analysis of *nodC*, ARDRA and rDNA sequence analysis. Soil Biol Biochem 2009; 41: 45-53.

[102]   Zhang YF, Wang ET, Tian CF, Qin F. *Bradyrhizobium elkanii, Bradyrhizobium yuanmingense* and *Bradyrhizobium japonicum* are the main rhizobia associated with *Vigna unguiculata* and *Vigna radiata* in the subtropical region of China. FEMS Microbiol Lett 2008; 285: 146-54.

[103] Hernandez-Lucas I, Segovia L, Martinez-Romero E, Pueppke SG. Phylogenetic relationships and host range of *Rhizobium* spp. that nodulate *Phaseolus vulgaris* L. Appl Envir Microbiol 1995; 61: 2775-79.

[104] Lindstrom K. *Rhizobium galegae*, a new species of legume root nodule bacteria. Int J Syst Bacteriol 1989; 39: 365-67.

[105] Lipsanen P, Lindstrom K. Lipopolysaccharide and Protein Patterns of *R. galegae*. In: Bothe H, de Bruijn FJ, Newton WE, Eds. Nitrogen Fixation: Hundred Years After. Stuttgart, VCH Publishers, 1988; p. 478.

[106] Wang ET, van Berkum P, Beyene D, *et al. Rhizobium huautlense* sp. nov., a symbiont of *Sesbania herbacea* that has a close phylogenetic relationship with *Rhizobium galegae*. Int J Syst Bacteriol 1998; 48: 687-99.

[107] Amarger N, Macheret V, Laguerre G. *Rhizobium gallicum* sp. nov. and *Rhizobium giardinii* sp. nov. from *Phaseolus vulgaris* nodules. Int J Syst Bacteriol 1997; 47: 996-06.

[108] Lopez-Lopez A, Rogel-Hernnndez MA, Boris I, *et al. Rhizobium grahamii* sp. nov. from *Dalea leporina, Leucaena leucocephala, Clitoria ternatea* nodules and *Rhizobium mesoamericanum* sp. nov. from *Phaseolus vulgaris*, siratro, cowpea and *Mimosa pudica* nodules. Int J Syst Evol Microbiol 2011; doi: 10.1099/ijs.0.033555-0.

[109] Chen WX, Tan ZY, Gao JL, *et al. Rhizobium hainanense* sp. nov. isolated from tropical legumes. Int J Syst Bacteriol 1997; 47: 870-73.

[110] Ren DW, Wang ET, Chen WF, *et al. Rhizobium herbae* sp. nov. and *Rhizobium giardinii*-related bacteria, minor microsymbionts of various wild legumes in China. Int J Syst Evol Microbiol 2011; 61: 1912-20.

[111] Ramirez-Bahena MH, Garcia-Fraile P, Peix A, *et al.* Revision of the taxonomic status of the species *Rhizobium leguminosarum* (Frank 1879) Frank 1889[AL], *Rhizobium phaseoli* Dangeard 1926[AL] and *Rhizobium trifolii* Dangeard 1926[AL]. *R. trifolii* is a later synonym of *R. leguminosarum*. Reclassification of the strain *R. leguminosarum* DSM 30132 (=NCIMB 11478) as *Rhizobium pisi* sp. nov. Int J Syst Evol Microbiol 2008; 58: 2484-90.

[112] Oyaizu H, Naruhashi N, Gamou T. Molecular methods of analysing bacterial diversity: the case of rhizobia. Biodivers Conserv 1992; 1: 237-49.

[113] Allen ON, Allen EK. The Leguminosae. A Source Book of Characteristics, Uses and Nodulation. University of Wisconsin Press, Madison, USA. 1981.

[114] Trinick MJ, Goodchild DJ, Miller C. Localization of bacteria and hemoglobin in root nodules of *Parasponia andersonii* containing both *Bradyrhizobium* strains and *Rhizobium leguminosarum* biovar *trifolii*. Appl Envir Microbiol 1989; 55: 2046-55.

[115] Laguerre G, Fernandez MP, Edel V, *et al.* Genomic heterogeneity among French *Rhizobium* strains isolated from *Phaseolus vulgaris* L. Int J Syst Bacteriol 1993; 43: 761-67.

[116] Martinez-Romero E, Segovia L, Mercante FM, *et al. Rhizobium tropici*, a novel species nodulating *Phaseolus vulgaris* L. beans and *Leucaena* sp. trees. Int J Syst Bacteriol 1991; 41: 417-26.

[117] Karunakaran R, Ramachandran, Seaman JC, *et al.* Transcriptomic analysis of *Rhizobium leguminosarum* biovar viciae in symbiosis with host plants *Pisum sativum* and *Vicia cracca*. J Bacteriol 2009; 191: 4002-14.

[118] Laguerre G, Louvrier P, Allard M, Amarger N. Compatibility of rhizobial genotypes within natural populations of *Rhizobium leguminosarum* biovar viciae for nodulation of host legumes. Appl Envir Microbiol 2003; 69: 2276-83.

[119] Almendras AS, Bottomley PJ. Influence of lime and phosphate on nodulation of soil-grown *Trifolium subterraneum* L. by indigenous *Rhizobium trifolii*. Appl Envir Microbiol 1987; 53: 2090-97.

[120] Zhang XX, Kosier B, Priefer UB. Symbiotic plasmid rearrangement in *Rhizobium leguminosarum* bv. viciae VF39SM. J Bacteriol 2001; 183: 2141-44.

[121] Ribeiro RA, Rogel MA, Lopez-Lopez A, *et al.* Reclassification of *Rhizobium tropici* type A strains as *Rhizobium leucaenae* sp. nov. Int J Syst Evol Microbiol 2011; doi: 10.1099/ijs.0.032912-0.

[122] Grandmaison J, Ibrahim R. Ultrastructural localization of a diprenylated isoflavone in *Rhizobium lupini-Lupinus albus* symbiotic association. J Exp Bot 1995; 46: 231-37.

[123] Valverde A, Igual JM, Peix A, *et al.* *Rhizobium lusitanum* sp. nov. a bacterium that nodulates *Phaseolus vulgaris*. Int J Syst Evol Microbiol 2006; 56: 2631-37.

[124] Lin DX, Chen WF, Wang FQ, *et al.* *Rhizobium mesosinicum* sp. nov., isolated from root nodules of three different legumes. Int J Syst Evol Microbiol 2009; 59: 1919-23.

[125] Gu CT, Wang ET, Tian CF, *et al.* *Rhizobium miluonense* sp. nov., a symbiotic bacterium isolated from *Lespedeza* root nodules. Int J Syst Evol Microbiol 2008; 58: 1364-68.

[126] Han TX, Wang ET, Wu JL, *et al.* *Rhizobium multihospitium* sp. nov., isolated from multiple legume species native of Xinjiang, China. Int J Syst Evol Microbiol 2008; 58: 1693-99.

[127] Pacovsky RS, Bayne HG, Bethlenfalvay GJ. Symbiotic interactions between strains of *Rhizobium phaseoli* and cultivars of *Phaseolus vulgaris* L. Crop Sci 1984; 24: 101-05.

[128] Panday D, Schumann P, Das SK. *Rhizobium pusense* sp. nov., isolated from the rhizosphere of chickpea (*Cicer arietinum* L.). Int J Syst Evol Microbiol 2011; 61: 2632-39.

[129] Squartini A, Struffi P, Doring H, *et al.* *Rhizobium sullae* sp. nov. (formerly '*Rhizobium hedysari*'), the root-nodule microsymbiont of *Hedysarum coronarium* L. Int J Syst Evol Microbiol 2002; 52: 1267-76.

[130] Hou BC, Wang ET, Li Y, Jr., *et al.* *Rhizobium tibeticum* sp. nov., a symbiotic bacterium isolated from *Trigonella archiducis-nicolai* (Sirj.) Vassilcz. Int J Syst Evol Microbiol 2009; 59: 3051-57.

[131] van Rhijn PJ, Feys B, Verreth C, Vanderleyden J. Multiple copies of *nodD* in *Rhizobium tropici* CIAT899 and BR816. J Bacteriol 1993; 175: 438-47.

[132] Amarger N, Bours M, Revoy F, *et al.* *Rhizobium tropici* nodulates field-grown *Phaseolus vulgaris* in France. Plant Soil 1994; 161: 147-56.

[133] Zhang RJ, Hou BC, Wang ET, *et al.* *Rhizobium tubonense* sp. nov., isolated from root nodules of *Oxytropis glabra*. Int J Syst Evol Microbiol 2011; 61: 512-17.

[134] Wang F, Wang ET, Wu LJ, *et al.* *Rhizobium vallis* sp. nov., isolated from nodules of three leguminous species. Int J Syst Evol Microbiol 2011; 61: 2582-88.

[135] Ren DW, Chen WF, Sui XH, *et al.* *Rhizobium vignae* sp. nov., a symbiotic bacterium isolated from multiple legume species. Int J Syst Evol Microbiol 2010; 61: 580-86.

[136] Tan ZY, F.L. FL, Peng GX, *et al.* *Rhizobium yanglingense* sp. nov., isolated from arid and semi-arid regions in China. Int J Syst Evol Microbiol 2001; 51: 909-14.

[137] Lopez-Lara IM, van der Drift KMGM, van Brussel AAN, *et al.* Induction of nodule primordia on *Phaseolus* and *Acacia* by lipo-chitin oligosaccharide nodulation signals from broad-host-range *Rhizobium* strain GRH2. Plant Mol Biol 1995; 29: 465-77.

[138] Ladha JK, So RB. Numerical taxonomy of photosynthetic rhizobia nodulating *Aeschynomene* species. Int J Syst Bacteriol 1994; 44: 62-73.

[139] Lorquin J, Lortet G, Ferro M, *et al.* *Sinorhizobium teranga* bv. acaciae ORS1073 and *Rhizobium* sp. strain ORS1001, two distantly related *Acacia*-nodulating strains, produce similar Nod factors that are O carbamoylated, N methylated and mainly sulfated. J Bacteriol 1997; 179: 3079-83.

[140] Rice DJ, Somasegaran P, MacGlashan K, Bohlool BB. Isolation of insertion sequence ISRLdTAL1145-1 from a *Rhizobium* sp. (*Leucaena diversifolia*) and distribution of homologous sequences identifying cross-inoculation group relationships. Appl Envir Microbiol 1994; 60: 4394-03.

[141]   Wilson JK. Leguminous Plants and their Associated Organisms. Cornell Univ Agric Exp Station Memoir 221. Cornell University Press, Ithaca, USA. 1939.

[142]   Pongsilp N, Leelahawonge C. Root-nodule symbionts of *Derris elliptica* Benth. are members of three distinct genera *Rhizobium, Sinorhizobium* and *Bradyrhizobium*. Int J Integr Biol 2010; 9: 37-42.

[143]   Estrella MJ, Munoz S, Soto MJ, *et al.* Genetic diversity and host range of rhizobia nodulating *Lotus tenuis* in typical soils of the Salado river basin (Argentina). Appl Envir Microbiol 2009; 75: 1088-98.

[144]   Eardly BD, Young JPW, Selander RK. Phylogenetic position of *Rhizobium* sp. strain Or 191, a symbiont of both *Medicago sativa* and *Phaseolus vulgaris*, based on partial sequences of the 16S rRNA and *nifH* genes. Appl Envir Microbiol 1992; 58: 1809-15.

[145]   Ogasawara M, Suzuki T, Mutoh I, *et al. Sinorhizobium indiaense* sp. nov. and *Sinorhizobium abri* sp. nov. isolated from tropical legumes, *Sesbania rostrata* and *Abrus precatorius*, respectively. Symbiosis 2003; 34: 53-68.

[146]   Toledo I, Lloret L, Martinez-Romero E. *Sinorhizobium americanus* sp. nov., a new *Sinorhizobium* species nodulating native *Acacia* spp. in Mexico. Syst Appl Microbiol 2003; 26: 54-64.

[147]   Young JM. The genus name *Ensifer* Casida 1982 takes priority over *Sinorhizobium* Chen *et al.* 1988 and *Sinorhizobium morelense* Wang *et al.* 2002 is a later synonym of *Ensifer adhaerens* Casida 1982. Is the combination '*Sinorhizobium adhaerens*' (Casida 1982) Willems *et al.* 2003 legitimate? Request for an Opinion. Int J Syst Evol Microbiol 2003; 53: 2107-10.

[148]   Nick G, de Lajudie P, Eardly BD, *et al. Sinorhizobium arboris* sp. nov. and *Sinorhizobium kostiense* sp. nov., isolated from leguminous trees in Sudan and Kenya. Int J Syst Bacteriol 1999; 49: 1359-68.

[149]   Rincon-Rosales R, Lloret L, Ponce E., Martinez-Romero E. Rhizobia with different symbiotic efficiencies nodulate *Acaciella angustissima* in Mexico, including *Sinorhizobium chiapanecum* sp. nov. which has common symbiotic genes with *Sinorhizobium mexicanum*. FEMS Microbiol Ecol 2009; 67: 103-17.

[150]   Saldana G, Martinez-Alcantara V, Vinardell JM, *et al.* Genetic diversity of fast-growing rhizobia that nodulate soybean (*Glycine max* L. Merr.). Arch Microbiol 2003; 180: 45-52.

[151]   Balatti PA, Pueppke SG. Nodulation of soybean by a transposon-mutant of *Rhizobium fredii* USDA 257 is subject to competitive nodulation blocking by other rhizobia. Plant Physiol 1990; 94: 1276-81.

[152]   Broughton WJ, Heycke N, Meyerz AH, Pankhurst CE. Plasmid-linked *nif* and *nod* genes in fast-growing rhizobia that nodulate *Glycine max, Psophocarpus tetragonolobus* and *Vigna unguiculata*. Proc Natl Acad Sci USA. 1984; 81: 3093-97.

[153]   Cregan PB, Keyser HH. Influence of *Glycine* spp. on competitiveness of *Bradyrhizobium japonicum* and *Rhizobium fredii*. Appl Envir Microbiol 1988; 54: 803-08.

[154]   Jiang G, Krishnan AH, Kim YW, *et al.* Functional *myo*-inositol dehydrogenase gene is required for efficient nitrogen fixation and competitiveness of *Sinorhizobium fredii* USDA 191 to nodulate soybean (*Glycine max* [L.] Merr.). J Bacteriol 2001; 183: 2595-04.

[155]   Scholla MH, Elkan GH. *Rhizobium fredii* sp. nov., a fast-growing species that effectively nodulates soybeans. Int J Syst Bacteriol 1984; 34: 484-86.

[156]   Trinick MJ. Relationships amongst the fast-growing rhizobia of *Lablab purpureus, Leucaena leucocephala, Mimosa* spp., *Acacia farnesiana* and *Sesbania grandiflora* and their affinities with other rhizobial groups. J Appl Bacteriol 1980; 49: 39-53.

[157]   Herrera-Cervera JA, Caballero-Mellado J, Laguerre G, *et al.* At least five rhizobial species nodulate *Phaseolus vulgaris* in a Spanish soil. FEMS Microbiol Ecol 1999; 30: 87-97.

[158]   Sadowsky MJ, Cregan PB, Keyser HH. Nodulation and nitrogen fixation efficacy of *Rhizobium fredii* with *Phaseolus vulgaris* genotypes. Appl Envir Microbiol 1988; 54: 1907-10.

[159] Merabet C, Martens M, Mahdhi M, *et al.* Multilocus sequence analysis of root nodule isolates from *Lotus arabicus* (Senegal), *Lotus creticus, Argyrolobium uniflorum* and *Medicago sativa* (Tunisia) and description of *Ensifer numidicus* sp. nov. and *Ensifer garamanticus* sp. nov. Int J Syst Evol Microbiol 2010, 60. 664-74.

[160] Rome S, Fernandez MP, Brunel B, *et al. Sinorhizobium medicae* sp. nov. isolated from annual *Medicago* spp. Int J Syst Bacteriol 1996; 46: 972-82.

[161] de Lajudie P, Willems A, Pot B, *et al.* Polyphasic taxonomy of rhizobia: emendation of the genus *Sinorhizobium* and description of *Sinorhizobium meliloti* comb. nov., *Sinorhizobium saheli* sp. nov. and *Sinorhizobium teranga* sp. nov. Int J Syst Bacteriol 1994; 44: 715-33.

[162] Gu CT, Wang ET, Sui XH, Chen WX. Diversity and geographical distribution of rhizobia associated with *Lespedeza* spp. in temperate and subtropical regions of China. Arch Microbiol 2007; 188: 355-65.

[163] Eardly BD, Materon LA, Smith NH, *et al.* Genetic structure of natural populations of the nitrogen-fixing bacterium *Rhizobium meliloti*. Appl Envir Microbiol 1990; 56: 187-94.

[164] Pellock BJ, Cheng HP, Walker GC. Alfalfa root nodule invasion efficiency is dependent on *Sinorhizobium meliloti* polysaccharides. J Bacteriol 2000; 182: 4310-18.

[165] Biondi EG, Tatti E, Comparini D, *et al.* Metabolic capacity of *Sinorhizobium* (*Ensifer*) *meliloti* strains as determined by phenotype microarray analysis. Appl Envir Microbiol 2009; 75: 5396-04.

[166] Lloret L, Ormeno-Orrillo E, Rincon R, *et al. Ensifer mexicanus* sp. nov. a new species nodulating *Acacia angustissima* (Mill.) Kuntze in Mexico. Syst Appl Microbiol 2007; 30: 280-90.

[167] Lorquin J, Lortet G, Ferro M, *et al.* Nod factors from *Sinorhizobium saheli* and *S. teranga* bv. *sesbaniae* are both arabinosylated and fucosylated, a structural feature specific to *Sesbania rostrata* symbionts. Mol Plant-Microbe Interact 1997; 10: 879-90.

[168] Li QQ, Wang ET, Chang YL, *et al. Ensifer sojae* sp. nov., isolated from root nodules of *Glycine max* grown in saline-alkaline soils. Int J Syst Evol Microbiol 2011; 61: 1981-88.

[169] Peng GX, Tan ZY, Wang ET, *et al.* Identification of isolates from soybean nodules in Xinjiang region as *Sinorhizobium xinjiangense* and genetic differentiation of *S. xinjiangense* from *Sinorhizobium fredii*. Int J Syst Evol Microbiol 2002; 52: 457-62.

[170] Lafay B, Bullier E, Burdon JJ. Bradyrhizobia isolated from root nodules of *Parasponia* (Ulmaceae) do not constitute a separate coherent lineage. Int J Syst Evol Microbiol 2006; 56: 1013-18.

[171] Vinuesa P, Leon-Barrios M, Silva C, *et al. Bradyrhizobium canariense* sp. nov., an acid-tolerant endosymbiont that nodulates endemic genistoid legumes (Papilionoideae: Genisteae) from the Canary Islands, along with *Bradyrhizobium japonicum* bv. genistearum, *Bradyrhizobium* genospecies alpha and *Bradyrhizobium* genospecies beta. Int J Syst Evol Microbiol 2005; 55: 569-75.

[172] Chahboune R, Carro L, Peix A, *et al. Bradyrhizobium cytisi* sp. nov., isolated from effective nodules of *Cytisus villosus*. Int J Syst Evol Microbiol 2011; 61: 2922-27.

[173] Kuykendall LD, Saxena B, Devine TE, Udell SE. Genetic diversity in *Bradyrhizobium japonicum* Jordan 1982 and a proposal for *Bradyrhizobium elkanii* sp. nov. Can J Microbiol 1992; 38: 501-05.

[174] Ramsubhag A, Umaharan P, Donawa A. Partial 16S rRNA gene sequence diversity and numerical taxonomy of slow growing pigeon pea (*Cajanus cajan* L. Millsp) nodulating rhizobia. FEMS Microbiol Lett 2002; 216: 139-44.

[175] Jordan DC. Transfer of *Rhizobium japonicum* Buchanan 1980 to *Bradyrhizobium japonicum* gen. nov., a genus of slow-growing, root nodule bacteria form leguminous plants. Int J Syst Bacteriol 1982; 32: 136-39.

[176] Diatloff A, Date RA. CB756 and the broad spectrum strain concept. *Rhizobium* Newsl 1978; 23: 17-19.

[177] Appunu C, N'Zoue A, Laguerre G. Genetic diversity of native bradyrhizobia isolated from soybeans (*Glycine max* L.) in different agricultural-ecological-climatic regions of India. Appl Envir Microbiol 2008; 74: 5991-96.

[178] Eskew DL, Jiang Q, Caetano-Anolles C, Cresshoff PM. Kinetics of nodule development in *Glycine soja.* Plant Physiol 1993; 103: 1139-45.

[179] Gueye F, Moulin L, Sylla S, *et al.* Genetic diversity and distribution of *Bradyrhizobium* and *Azorhizobium* strains associated with the herb legume *Zornia glochidiata* sampled from across Senegal. Syst Appl Microbiol 2009; 32: 387-99.

[180] Ramirez-Bahena MH, Peix A, Rivas R, *et al. Bradyrhizobium pachyrhizi* sp. nov. and *Bradyrhizobium jicamae* sp. nov., isolated from effective nodules of *Pachyrhizus erosus*. Int J Syst Evol Microbiol 2009; 59: 1929-34.

[181] Chang YL, Wang JY, Wang ET, *et al. Bradyrhizobium lablabi* sp. nov., isolated from effective nodules of *Lablab purpureus* and *Arachis hypogaea*. Int J Syst Evol Microbiol 2011; 61: 2496-02.

[182] Xu LM, Ge C, Cui Z, *et al. Bradyrhizobium liaoningense* sp. nov., isolated from the root nodules of soybeans. Int Syst Bacteriol 1995; 45: 706-11.

[183] Schulze J, Temple G, Temple SJ, *et al.* Nitrogen fixation by white lupin under phosphorus deficiency. Ann Bot 2006; 98: 731-40.

[184] Rodriguez-Echeverria S, Crisostomo JA, Freitas H. Genetic diversity of rhizobia associated with *Acacia longifolia* in two stages of invasion of coastal sand dunes. Appl Envir Microbiol 2007; 73: 5066-70.

[185] Rodriguez-Echeverria S, Perez-Fernandez MA, Vlaar S, Finan T. Analysis of the legume–rhizobia symbiosis in shrubs from central western Spain. J Appl Microbiol 2003; 95: 1367-74.

[186] Bergersen FJ, Turner GL, Amarger N, *et al.* Strain of *Rhizobium lupini* determines natural abundance of ^{15}N in root nodules of *Lupinus* spp. Soil Biol Biochem 1986; 18: 97-01.

[187] Pongsilp N, Nuntagij A. Selection and characterization of mungbean root nodule bacteria based on their growth and symbiotic ability in alkaline conditions. Suranaree J Sci Technol 2007; 14: 277-86.

[188] Wang FQ, Wang ET, Liu J, *et al. Mesorhizobium albiziae* sp. nov., a novel bacterium that nodulates *Albizia kalkora* in a subtropical region of China. Int J Syst Evol Microbiol 2007; 57: 1192-99.

[189] Lorite MJ, Munoz S, Olivares J, *et al.* Characterization of strains unlike *Mesorhizobium loti* that nodulate *Lotus* spp. in saline soils of Granada, Spain. Appl Envir Microbiol 2010; 76: 4019-26.

[190] Chen WM, Zhu WF, Bontemps C, *et al. Mesorhizobium alhagi* sp. nov., isolated from wild *Alhagi sparsifolia* in north-western China. Int J Syst Evol Microbiol 2010; 60: 958-62.

[191] Nandasena KG, O'Hara GW, Tiwari RP, *et al. Mesorhizobium australicum* sp. nov. and *Mesorhizobium opportunistum* sp. nov., isolated from *Biserrula pelecinus* L. in Australia. Int J Syst Evol Microbiol 2009; 59: 2140-47.

[192] Chen WM, Zhu WF, Bontemps C, *et al. Mesorhizobium camelthorni* sp. nov., isolated from *Alhagi sparsifolia*. Int J Syst Evol Microbiol 2011; 61: 574-79.

[193] Guan SH, Chen WF, Wang ET, *et al. Mesorhizobium caraganae* sp. nov., a novel rhizobial species nodulated with *Caragana* spp. in China. Int J Syst Evol Microbiol 2008; 58: 2646-53.

[194] Velazquez E, Igual JM, Willems A, *et al. Mesorhizobium chacoense* sp. nov. a novel species that nodulates *Prosopis alba* in the Chaco Arido region (Argentina). Int J Syst Evol Microbiol 2001; 51: 1011-21.

[195] Nandasena KG, O'Hara GW, Tiwari RP, *et al. Mesorhizobium ciceri* biovar biserrulae, a novel biovar nodulating the pasture legume *Biserrula pelecinus* L. Int J Syst Evol Microbiol 2007; 57: 1041-45.

[196] Nour SM, Cleyet-Marel JC, Beck D, *et al.* Genotypic and phenotypic diversity of *Rhizobium* isolated from chickpea (*Cicer arietinum* L.). Can J Microbiol 1994; 40: 345-54.

[197] Han TX, Han LL, Wu LJ, *et al. Mesorhizobium gobiense* sp. nov. and *Mesorhizobium tarimense* sp. nov., isolated from wild legumes growing in desert soils of Xinjiang, China. Int J Syst Evol Microbiol 2008; 58: 2610-18.

[198] Chen WX, Li GS, Qi YL, *et al. Rhizobium huakuii* sp. nov. isolated from the root nodules of *Astragalus sinicus.* Int J Syst Bacteriol 1991; 41: 275-80.

[199] Yang GP, Debelle F, Savagnac A, *et al.* Structure of the *Mesorhizobium huakuii* and *Rhizobium galegae* Nod factors: a cluster of phylogenetically related legumes are nodulated by rhizobia producing Nod factors with alpha, beta-unsaturated N-acyl substitutions. Mol Microbiol 1999; 34: 227-37.

[200] de Lajudie P, Willems A, Nick G, *et al.* Characterization of tropical tree rhizobia and description of *Mesorhizobium plurifarium* sp. nov. Int J Syst Evol Microbiol 1998; 48: 369-82.

[201] Jarvis BDW, Pankhurst CE, Patel JJ. *Rhizobium loti*, a new species of legume root nodule bacteria. Int J Syst Bacteriol 1982; 32: 378-80.

[202] Saeki K, Kouchi H. The *Lotus* symbiont, *Mesorhizobium loti*: molecular genetic techniques and application. J Plant Res 2000; 113: 457-65.

[203] Vidal C, Chaintreuil C, Berge O, *et al. Mesorhizobium metallidurans* sp. nov., a metal-resistant symbiont of *Anthyllis vulneraria* growing on metallicolous soil in Languedoc, France. Int J Syst Evol Microbiol 2009; 59: 850-55.

[204] Wang ET, Kan FL, Tan ZY, *et al.* Diverse *Mesorhizobium plurifarium* populations native to Mexican soils. Arch Microbiol 2003; 180: 444-54.

[205] Zhou PF, Chen WM, Wei GH. *Mesorhizobium robiniae* sp. nov., isolated from root nodules of *Robinia pseudoacacia.* Int J Syst Evol Microbiol 2010; 60: 2552-56.

[206] Gao JL, Turner SL, Kan FL, *et al. Mesorhizobium septentrionale* sp. nov. and *Mesorhizobium temperatum* sp. nov., isolated from *Astragalus adsurgens* growing in the northern regions of China. Int J Syst Evol Microbiol 2004; 5: 2003-12.

[207] Lu YL, Chen WF, Wang ET, *et al. Mesorhizobium shangrilense* sp. nov., isolated from root nodules of *Caragana* species. Int J Syst Evol Microbiol 2009; 59: 3012-18.

[208] Chen W, Wang E, Wang S, *et al.* Characteristics of *Rhizobium tianshanense* sp. nov., a moderately and slowly growing root nodule bacterium isolated from an arid saline environment in Xinjiang, People's Republic of China. Int J Syst Bacteriol 1995; 45: 153-59.

[209] Tan ZY, Xu XD, Wang ET, *et al.* Phylogenetic and genetic relationships of *Mesorhizobium tianshanense* and related rhizobia. Int J Syst Bacteriol 1997; 47: 874-79.

[210] Rivas R, Peix A, Mateos PF, *et al.* Biodiversity of populations of phosphate solubilizing rhizobia that nodulates chickpea in different Spanish soils. In: Velazquez E, Rodriguez-Barrueco C, Eds. First International Meeting on Microbial Phosphate Solubilization. Dordrecht, Springer, 2007; pp. 23-33.

[211] Lee KB, de Backer P, Aono T, *et al.* The genome of the versatile nitrogen fixer *Azorhizobium caulinodans* ORS571. BMC Genomics 2008; 9: 271.

[212] Vandamme P, Coenye T. Taxonomy of the genus *Cupriavidus*: a tale of lost and found. Int J Syst Evol Microbiol 2004; 54: 2285-89.

[213] Barrett CF, Parker MA. Prevalence of *Burkholderia* sp. nodule symbionts on four mimosoid legumes from Barro Colorado Island, Panama. Syst Appl Microbiol 2005; 28: 57-65.

CHAPTER 2

Phenotypic Diversity of Rhizobia Assessed by Numerical Analysis, Enzyme Pattern and Serological Study

Neelawan Pongsilp[*]

Department of Microbiology, Faculty of Science, Silpakorn University, Nakhon Pathom, Thailand

Abstract: Phenotypic diversity of rhizobia has been studied by several methods, particularly numerical analysis, enzyme pattern and serological study. Numerical analysis employs a large range of biochemical and metabolic tests to differentiate among rhizobial species. The results obtained from numerical analysis support the proposal of several novel species of rhizobia. Rhizobial species vary in their enzymatic production and several enzymes are found to be necessary for the symbiotic effectiveness. Serological techniques are the most specific methods for identifying rhizobia based on natural marker characteristics. Phenotypic characterization by serotyping has been used to explore rhizobial populations in different geographical origins. Rhizobia in different serogroups have been found predominant among field populations.

Keywords: Antibody, Antigen, Biochemical characteristic, Enzyme pattern, Numerical analysis, Phenotypic diversity, Serocluster, Serogroup, Serological study, Rhizobia.

2.1. NUMERICAL ANALYSIS OF RHIZOBIA

Numerical analysis has been used widely to study phenotypic diversity of rhizobia. Many previous studies have used a large range of biochemical and metabolic tests to differentiate among rhizobial species. The phenotypic characteristics used frequently in numerical analysis are i) utilization of carbon sources such as adonitol, ammonium oxalate, arabinose, arabitol, calcium malonate, cascin hydrolysate, cellobiose, creatinine, dulcitol, esculin, ethylene glycol, fructose, galactose, glucose, glutamine, glycerol, inositol, inulin, isopropanol, lactose, lyxose, maltose, mannitol, mannose, melezitose, melibiose, meso-erythritol, salicin, sodium citrate, sodium gluconate, sodium hippurate,

---

*Address correspondence to Neelawan Pongsilp: Department of Microbiology, Faculty of Science, Silpakorn University, Nakhon Pathom, Thailand; Tel: +66-34-245337; Fax: +66-34-245336-37; E-mail: neelawan@su.ac.th

sodium lactate, sodium malate, sodium pyruvate, sodium succinate, sorbitol, sorbose, sucrose, raffinose, rhamnose, ribose, tagatose, trehalose, turanose, vanillic acid and xylose; ii) utilization of nitrogen sources such as alanine, arginine, aspartic acid, cysteine, cytosine, 6-furfurylaminopurine, glutamic acid, glycine, histidine, isoleucine, leucine, lysine, methionine, phenylalanine, proline, serine, threonine, thymine, tryptophan, tyrosine and valine; iii) requirement for vitamins such as aminobenzoic acid (vitamin B), ascorbic acid (vitamin C), biotin (vitamin B7), calcium-pantothenate (vitamin B5), cyanocobalamin (vitamin B12), folic acid (vitamin B9), myo-inositol (vitamin B8), nicotinic acid or nicotinamide (vitamin B3), pyridoxine hydrochloride (vitamin B6), riboflavin (vitamin B2), thiamine hydrochloride (vitamin B1) and thioctic acid (vitamin N); iv) tolerance to dyes such as auramine O.B.S, bismarck brown, brilliant cresyl blue, bromocresol purple, bromophenol blue, bromthymol blue, chrysoidine, congo red, erythrosin A, fuchsin basic, gentian violet, giemsa stain, iodine green, light green SF, methylene blue, methyl green, methyl red, methyl violet, neutral red, nigrosine, nile blue, orcein, phenol red, picrocarmine, rosanilin, rose bengal, safranine T and sudan 1; v) tolerance to antibiotics such as ampicillin, cefotaxine, ceftazdime, chloramphenicol, doxycycline, erythromycin, gentamycin, kanamycin, medemycin, novobiocin, penicillin, spectinomycin, streptomycin, terramycin, tetracycline, ureomycin and vancomycin; vi) tolerance to pesticides such as carbendazol, chlordimeform, dimethoate, malathion, omethoate and phoxin; vii) growth rate; viii) growth at different pH; ix) growth at different temperatures; x) growth at different NaCl concentrations; xi) growth in peptone broth; xii) reaction in litmus milk; xiii) reduction of methyl blue; xiv) reduction of nitrate; xv) production of enzymes such as catalase, cytochrome oxidase, gelatinase, peroxidase, phenylalaninase and urease; xvi) production of acid or alkaline; xvii) colony morphology; xviii) the presence of peritrichous or polar flagella [1-8].

Biochemical characteristics including production of gelatinase, tolerance to 2% NaCl, acid and alkaline were found to distinguish the fast- and slow-growing soybean rhizobia [7]. According to Xu *et al.* [8], the extra-slow growing (ESG) soybean rhizobia were compared with reference strains belonging to the genera *Agrobacterium*, *Bradyrhizobium* and *Rhizobium* by performing a numerical analysis of 191 phenotypic features. *Bradyrhizobium* and *Rhizobium* strains were placed in 2 distinct phenotypic groups. All of the ESG strains examined clustered closely in the genus *Bradyrhizobium* but were separated from *Bradyrhizobium japonicum* at the species level. On the basis of numerical analysis and techniques based on DNA analysis including guanine-cytosine (GC) content, DNA-DNA hybridization, small subunit ribosomal RNA gene (16S rDNA) sequence analysis and carbon:nitrogen (C:N) ratio analysis, the ESG soybean rhizobia were proposed for the new species, *Bradyrhizobium liaoningense*. Chen *et al.* [2] performed a numerical analysis of 148 phenotypic characteristics of rhizobia isolated from an acrid saline desert soil. The results obtained from numerical analysis and DNA homology values supported that the moderate- and slow-growing strains which produced acid are members of the new species, *Rhizobium tianshanense*. Chen *et al.* [3] proposed *Rhizobium hainanense* on the basis of 16S rRNA gene sequencing, DNA-DNA hybridization and phenotypic characterization. *Rhizobium multihospitium* was proposed according to phenotypic characteristics that differentiated it from closely related *Rhizobium* species including *R. etli*, *R. leguminosarum*, *R. lusitanum*, *R. rhizogenes* (currently *Agrobacterium rhizogenes*) and *R. tropici*. The differentiating characteristics included utilization of sodium formate as a sole carbon source, utilization of L-threonine and D-threonine as sole nitrogen sources as well as resistance to chloramphenicol (100 µg/ml) and erythromycin (100 µg/ml) [9]. Lin *et al.* [10] proposed a novel species of rhizobia, *Shinella kummerowiae*, as the type strain CCBAU 25048^T could be differentiated from the 2 defined *Shinella* species based on several phenotypic characteristics including the use of citrate and D-ribose as sole carbon sources, the growth at pH 11.0 as well as the fatty acid composition. Substrate utilization can be performed using the Phenotype Micro Array (PM) system (Biolog) (Biolog Inc., CA). Hundreds of different growth conditions were tested in order to compare the metabolic capabilities of the reference strain *Ensifer meliloti* (formerly *Sinorhizobium meliloti*) Rm1021 with natural *E. meliloti* isolates. The results of PM analysis showed that most phenotypic differences involved utilization of carbon sources as well as tolerance to osmolytes and pH, while fewer differences were scored for utilization of nitrogen, phosphorus and sulfur sources [11]. The taxonomic relationship made by computer-assisted numerical taxonomy can be presented as a dendrogram. The dendrogram can be constructed using similarity (S_J) and matching (S_{SM}) coefficients and the unweighted pair group method with arithmetic mean (UPGMA) [12]. The dendrogram constructed from 56 phenotypic features (growth characteristics, utilization of carbon and nitrogen sources as well as intrinsic antibiotic resistance) showed that the rhizobial strains from root nodules of temperate-zone legumes formed a cluster separate from both *Bradyrhizobium* and *Rhizobium* [13].

However, the use of traditional methods has been limited mainly by highly similar phenotypic characteristics that usually occur among closely related strains and by diverse phenotypic characteristics that usually occur among distantly related strains. Pongsilp *et al.* [6] found that *Rhizobium* and *Bradyrhizobium* that nodulate *Pueraria mirifica* (white Kwao Kruea) exhibited diverse phenotypic features. Strains belonging to the same genera varied in their phenotypic features including utilization of carbon and nitrogen sources, resistance to antibiotics, requirement for vitamins and production of enzymes. Both genera could not be distinguished from one another based on these characteristics. This finding suggests that bacteria in different genera may adapt to the environmental conditions influenced by root exudates from their host. As root exudates are composed of both low and high molecular weight components, including an array of primary and secondary metabolites, proteins and peptides [14, 15], that vary in quantity and chemical structure depending on the plant species, therefore root exudates can provide the selective environments for specific groups of bacteria.

2.2. ENZYME PATTERN OF RHIZOBIA

Nitrogenase enzyme involved in nitrogen fixation and chitin synthase for synthesis of Nod factors are key enzymes found in every member of rhizobia. Besides enzymes involved in symbiotic properties, cellulolytic, hydrolytic, metabolic and pectinolytic enzymes have been detected in many rhizobial strains. Variability in hydrolytic enzymes may be related to variations in host plant specificity of rhizobia [16]. The differences in enzyme types and activities found among the fast- and slow-growing strains may reflect the differences in infection pathways [17]. Cellulase activity has been detected in a wide range of rhizobia, indicating that the production of plant cell wall degrading enzymes is a general characteristic of rhizobia [18]. It clearly seems that amylase, cellulase, glucosidase, maltase and trehalase are necessary for the establishment of nodule symbiosis [19]. Several, but not all enzymes in carbohydrate catabolism are involved in symbiotic properties. The possession of a hydrogenase uptake enzyme in *B. japonicum* (formerly *Rhizobium japonicum*) increased the efficiency of nitrogen fixation, with a concomitant increase in plant weight and nitrogen content [20-22]. Aminotransferase required for aspartate catabolism was found to be essential for symbiotic nitrogen fixation of *E. meliloti* (formerly *Rhizobium meliloti*), implicating that aspartate is an essential substrate for bacteria in the nodule [23]. *E. meliloti* also required an active acetohydroxy acid isomeroreductase enzyme, the second enzyme in the parallel pathways for the biosynthesis of isoleucine and valine, for successful nodulation of *Medicago sativa* (alfalfa) [24]. On the contary, fructokinase, glucokinase and pyruvate dehydrogenase were not necessary for *R. leguminosarum* to nodulate and fix nitrogen with *Pisum sativum* (pea), therefore the capacity to utilize particular C_6 and C_{12} sugars is apparently not essential for the bacteroid development or the establishment of effective nitrogen fixation [25]. *Rhizobium trifolii* mutants, defective in the production of galactose dehydrogenase, glucokinase and pyruvate carboxylase, formed an effective symbiosis on *Trifolium pratense* (red clover), suggesting that neither fructose, glucose nor sucrose are used by the bacteroids to provide adenosine triphosphate (ATP) and reductant for nitrogen fixation [26]. Besides the enzymes shared in common, an enzymatic differentiation was found among different genera of rhizobia. A nicotinamide adenine dinucleotide phosphate-6-phosphogluconate dehydrogenase was detected in strains of the fast-growing group but not in strains of the slow-growing group [27]. Examples of enzymes produced by rhizobia are shown in Table **1**.

Table 1: Examples of enzymes produced by rhizobia

Enzyme	Enzyme Function	Rhizobial Species	References
ACC deaminase	degrades1-aminocyclopropane-1-carboxylate (ACC), a substrate for ethylene synthesis	*Mesorhizobium loti* *Rhizobium leguminosarum* *Rhizobium* spp.	[28-31]
acetoacetate-succinyl-CoA transferase	catalyzes the chemical reaction between succinyl-CoA and 3-oxo acid, releasing succinate and 3-oxoacyl-CoA	*Bradyrhizobium japonicum* (formerly *Rhizobium japonicum*)	[32]

Table 1: cont....

acetohydroxy acid isomeroreductase (isomeroreductase)	synthesizes isoleucine and valine	*Ensifer meliloti* (formerly *Rhizobium meliloti*)	[24]
N-acetyl-β-glucosaminidase	degrades murein (peptidoglycan)	*Bradyrhizobium* spp.	[6]
acid phosphatase	removes a phosphate group from an organic compound at acid condition	*Bradyrhizobium* spp. *Cupriavidus* spp. *Mesorhizobium* spp. *Rhizobium* spp. *Sinorhizobium* spp.	[6, 33, 34]
aconitase	catalyzes the isomerization of citrate to isocitrate in the tricarboxylic acid cycle	*Rhizobium etli* *Rhizobium tropici* *Rhizobium* spp.	[35]
adenylate kinase	catalyzes the interconversion of adenine	*Ensifer saheli* (formerly *Sinorhizobium saheli*) *Ensifer terangae* (formerly *Sinorhizobium terangae*) *Rhizobium galegae* *Rhizobium huautlense*	[36]
alanine dehydrogenase	catalyzes the chemical reaction between L-alanine and nicotinamide adenine dinucleotide (NAD^+), releasing pyruvate, ammonia, NADH and H^+	*E. terangae* *R. etli* *R. galegae* *R. huautlense* *R. tropici* *Rhizobium* spp.	[35, 36]
alcohol dehydrogenase (ADH)	catalyzes the oxidation of alcohol	*E. meliloti*	[37]
aldolase	degrades aldol	*E. meliloti*	[37]
alkaline phosphatase	removes a phosphate group from an organic compound at alkaline condition	*Bradyrhizobium* spp. *Cupriavidus* spp. *Mesorhizobium* spp. *Rhizobium* spp. *Sinorhizobium* spp.	[6, 33, 34]
δ-aminolevulinic acid synthetase	synthesizes heme	*E. meliloti*	[38]
aminotransferase, aspartate amino-transferase (AAT)	catalyzes the transfer of an amino group from an amino acid to an alpha-keto acid	*E. meliloti*	[23, 39]
α- and β-arabinosidase	degrades arabinose	*Rhizobium lusitanum*	[40]

Table 1: cont….

asparaginase	degrades asparagine	*R. etli*	[41]
aspartate ammonialyase (L-aspartase)	degrades asparate	*R. etli* *R. leguminosarum*	[41, 42]
β-carboxy-*cis,cis*-muconate lactonizing enzyme (CMLE)	catalyzes the reversible gamma-lactonization of 3-carboxy-*cis, cis*-muconate	*B. japonicum* *Bradyrhizobium* spp. *Ensifer fredii* (formerly *Rhizobium fredii*) *E. meliloti* *R. leguminosarum* *Rhizobium trifolii*	[43]
carboxymethyl celluase (CMCase)	degrades carboxymethyl cellulose	*R. leguminosarum* *Rhizobium* sp.	[44, 45]
γ-carboxymucono-lactone decarboxylase	converts 2-carboxy-2,5-dihydro-5-oxofuran-2-acetate to 4,5-dihydro-5-oxofuran-2-acetate and carbon dioxide	*B. japonicum* *Bradyrhizobium* spp. *E. fredii* *E. meliloti* *R. leguminosarum* *R. trifolii*	[43]
catalase	catalyzes the dismutation of hydrogen peroxide	*Bradyrhizobium yuanmingense* *Devosia neptuniae*	[46, 47]
cellobiase	degrades cellobiose	*R. lusitanum*	[40]
cellulase	degrades cellulose	*R. leguminosarum* *R. trifolii* *Rhizobium* sp.	[16, 48]
α-chymotrypsin	degrades peptide	*Bradyrhizobium* spp. *Rhizobium* spp.	[6]
cystine acrylamidase	degrades cystine-containing peptide	*Bradyrhizobium* spp. *Rhizobium* spp.	[6]
esterase	degrades ester	*Azorhizobium caulinodans* *Bradyrhizobium elkanii* *B. japonicum* *Bradyrhizobium* spp. *E. fredii* *E. meliloti* *E. saheli* *E. terangae* *Mesorhizobium amorphae*	[6, 35, 49]

Table 1: cont....

		Mesorhizobium ciceri	
		Mesorhizobium huakuii	
		Mesorhizobium loti	
		Mesorhizobium mediterraneum	
		Mesorhizobium plurifarium	
		Mesorhizobium tianshanense	
		Mesorhizobium spp.	
		R. etli	
		R. galegae	
		R. leguminosarum	
		R. tropici	
		Rhizobium spp.	
ferrireductase	reduces ferric ion (Fe^{3+})	*Rhizobium* sp.	[50]
fructokinase	catalyzes the transfer of a phosphate group from adenosine triphosphate (ATP) to fructose	*R. leguminosarum*	[25, 51]
fructose bisphosphatase	converts fructose-1,6-bisphosphate to fructose 6-phosphate in gluconeogenesis pathway	*R. tropici*	[52]
fructose-6-phosphate aldolase	catalyzes the aldol cleavage of fructose 6-phosphate	*R. tropici*	[52]
galactose dehydrogenase	catalyzes the reaction between D-galactose and NAD^+, releasing D-galactono-1,4-lactone, NADH and H^+	*R. trifolii*	[26]
α-galactosidase	degrades β-galactosides such as lactose	*R. lusitanum*	[40]
β-galactosidase	degrades β-galactosides such as lactose	*D. neptuniae* *E. meliloti* *Ochrobactrum cytisi* *R. lusitanum* *R. trifolii*	[40, 46, 53-55]
glucokinase	adds a phosphate group to glucose (glucose phosphorylation)	*R. leguminosarum* *R. trifolii* *R. tropici*	[25, 26, 52]
glucose-6-phosphate dehydrogenase	converts glucose-6-phosphate into 6-phosphoglucono-δ-lactone in the pentose phosphate pathway	*A. caulinodans* *B. elkanii* *B. japonicum* *Bradyrhizobium* spp. *E. fredii* *E. meliloti* *E. saheli* *E. terangae*	[35-37, 49, 52]

Table 1: cont….

		M. amorphae	
		M. ciceri	
		M. huakuii	
		M. loti	
		M. mediterraneum	
		M. plurifarium	
		M. tianshanense	
		Mesorhizobium spp.	
		R. etli	
		R. galegae	
		R. huautlense	
		R. leguminosarum	
		R. tropici	
		Rhizobium spp.	
α-glucosidase	degrades glucosides	*Bradyrhizobium* spp.	[6, 19]
		Rhizobium spp.	
β-glucosidase	degrades glucosides	*Bradyrhizobium* spp.	[6, 53, 55]
		O. cytisi	
		R. trifolii	
glutamate dehydrogenase	converts glutamate to α-ketoglutarate and vice versa	*R. etli*	[35]
		R. tropici	
		Rhizobium spp.	
glutaminase	degrades glutamine	*R. etli*	[56]
glutamine transaminase	catalyzes the reaction between L-glutamine and phenylpyruvate, releasing 2-oxoglutaramate and L-phenylalanine	*R. etli*	[56]
glycanase	degrades carboxymethyl cellulose and exopolysaccharide	*R. leguminosarum*	[57]
hemicellulase	degrades hemicellulose	*R. trifolii*	[16, 58]
		Rhizobium spp.	
hexokinase	adds a phosphate group to hexose such as glucose and galactose (phosphorylation)	*A. caulinodans*	[35, 36]
		B. elkanii	
		B. japonicum	
		Bradyrhizobium spp.	
		E. fredii	
		E. meliloti	
		E. saheli	
		E. terangae	
		M. amorphae	
		R. huautlense	
		R. leguminosarum	
		R. tropici	
		Rhizobium spp.	

Table 1: cont....

hydrogenase	oxidizes dihydrogen (H_2)	*B. japonicum*	[20, 59, 60]
4-hydroxybenzoate hydroxylase	converts hydroxybenzoate into hydroquinone	*R. leguminosarum*	[61]
hydroxybutyrate dehydrogenase	converts 3-hydroxybutanoate into acetoacetate	*B. japonicum*	[32]
indolphenol oxidase	catalyzes the transport of electrons from NADH to electron acceptors (usually oxygen)	*A. caulinodans* *B. elkanii* *B. japonicum* *Bradyrhizobium* spp. *E. fredii* *E. meliloti* *E. saheli* *E. terangae* *M. amorphae* *M. ciceri* *M. huakuii* *M. loti* *M. mediterraneum* *M. plurifarium* *M. tianshanense* *Mesorhizobium* spp. *R. etli* *R. galegae* *R. leguminosarum* *R. tropici* *Rhizobium* spp.	[35, 49]
isocitrate dehydrogenase	catalyzes the oxidative decarboxylation of isocitrate in the citric acid cycle	*A. caulinodans* *B. elkanii* *B. japonicum* *Bradyrhizobium* spp. *E. fredii* *E. meliloti* *E. saheli* *E. terangae* *M. amorphae* *M. ciceri* *M. huakuii* *M. loti* *M. mediterraneum* *M. plurifarium* *M. tianshanense* *Mesorhizobium* spp.	[32, 35-37, 49]

Table 1: cont....

		R. etli R. galegae R. huautlense R. leguminosarum R. tropici Rhizobium spp.	
β-ketoadipate enol-lactone hydrolase	converts β-ketoadipate enol-lactone to β-ketoadipate	B. japonicum Bradyrhizobium spp E. fredii E. meliloti R. leguminosarum R. trifolii	[43]
β-ketoadipate succinyl-coenzyme A transferase	catalyzes the transfer of CoA thioester bond from succinate to β-ketoadipate	B. japonicum Bradyrhizobium spp E. fredii E. meliloti R. leguminosarum R. trifolii	[43]
β-ketothiolase	condenses 2 molecules of acetyl coenzyme A (acetyl-CoA) to acetoacetyl-CoA.	B. japonicum	[32]
leucine acrylamidase	degrades leucine-containing peptide	Bradyrhizobium spp. Rhizobium spp.	[6]
malate dehydrogenase	converts malate into oxaloacetate and vice versa	B. japonicum E. meliloti E. saheli E. terangae R. etli R. galegae R. huautlense R. tropici Rhizobium spp.	[32, 35-37]
α-maltosidase	degrades phehyl α-maltoside	R. lusitanum	[40]
NADP-dependent glutamate dehydrogenase	catalyzes the reversible amination of 2-oxoglutarate to form glutamate	A. caulinodans B. elkanii B. japonicum Bradyrhizobium spp. E. fredii E. meliloti E. saheli E. terangae M. amorphae M. ciceri	[49]

Table 1: cont....

		M. huakuii	
		M. loti	
		M. mediterraneum	
		M. plurifarium	
		M. tianshanense	
		Mesorhizobium spp.	
		R. etli	
		R. galegae	
		R. huautlense	
		R. leguminosarum	
		R. tropici	
		Rhizobium spp.	
NAD-dependent phosphogluconate dehydrogenase	converts 6- phosphogluconate to ribulose 5-phosphate in the pentose phosphate pathway	R. tropici	[52]
NADP-malate dehydrogenase (malic enzyme)	catalyzes the chemical reaction between L-malate and nicotinamide adenine dinucleotide phosphate ($NADP^+$), releasing pyruvate, carbon dioxide and NADPH	A. caulinodans B. elkanii B. japonicum Bradyrhizobium spp. E. fredii E. meliloti E. saheli E. terangae M. amorphae M. ciceri M. huakuii M. loti M. mediterraneum M. plurifarium M. tianshanense Mesorhizobium spp. R. etli R. galegae R. huautlense R. leguminosarum R. tropici Rhizobium spp.	[36, 49, 62]
napthol-AS-BI-phosphohydrolase	removes a phosphate group from an organic compound	Bradyrhizobium spp. Rhizobium spp.	[6]
oxidase	catalyzes the transport of electrons from NADH to electron acceptors (usually oxygen)	D. neptuniae	[46]
pectinase	degrades pectin	R. etli Rhizobium sp.	[18, 63]

Table 1: cont….

phosphoenol pyruvate carboxy kinase	converts oxaloacetate into phosphoenolpyruvate (PEP) and carbon dioxide in gluconeogenesis pathway	*R. tropici*	[52]
phosphoglucomu-tase	transfers a phosphate group on an α-D-glucose monomer	*A. caulinodans* *B. elkanii* *B. japonicum* *Bradyrhizobium* spp. *E. fredii* *E. meliloti* *E. saheli* *E. terangae* *M. amorphae* *M. ciceri* *M. huakuii* *M. loti* *M. mediterraneum* *M. plurifarium* *M. tianshanense* *Mesorhizobium* spp. *R. etli* *R. galegae* *R. huautlense* *R. leguminosarum* *R. tropici* *Rhizobium* spp.	[35, 36, 49, 52]
6-phosphogluconate dehydrogenase	catalyzes the decarboxylation of 6-phosphogluconate into ribulose 5-phosphate in the pentose phosphate pathway	*E. meliloti*	[37]
phosphoglucose isomerase	catalyzes the interconversion of D-glucose 6-phosphate and D-fructose 6-phosphate	*A. caulinodans* *B. elkanii* *B. japonicum* *Bradyrhizobium* spp. *E. fredii* *E. meliloti* *E. saheli* *E. terangae* *M. amorphae* *M. ciceri* *M. huakuii* *M. loti* *M. mediterraneum* *M. plurifarium* *M. tianshanense*	[36, 49]

Table 1: cont....

		Mesorhizobium spp. R. etli R. galegae R. huautlense R. leguminosarum R. tropici Rhizobium spp.	
phosphotransferase	adds a phosphate group to an organic molecule	R. leguminosarum	[64]
polygalacturonase	degrades polygalacturonan	E. meliloti R. leguminosarum	[44, 65]
protocatechuate oxygenase	converts protocatechuate to 4-carboxy-2-hydroxymuconate semialdehyde	B. japonicum Bradyrhizobium spp. E. fredii E. meliloti R. leguminosarum R. trifolii	[43]
pyruvate carboxylase	catalyzes the carboxylation of pyruvate to form oxaloacetate	R. trifolii	[26]
pyruvate dehydrogenase	converts pyruvate into acetyl-CoA	B. japonicum R. leguminosarum	[25, 32]
pyruvate kinase	catalyzes the transfer of a phosphate group from phosphoenolpyruvate (PEP) to adenosine diphosphate (ADP), releasing pyruvate and ATP	E. meliloti	[37]
pyruvate orthophosphate dikinase	catalyzes the chemical reaction of ATP, pyruvate and phosphate, releasing adenosine monophosphate (AMP), phosphoenolpyruvate and diphosphate	E. meliloti	[66]
trypsin	degrades peptide	Bradyrhizobium spp. Rhizobium spp.	[6]
UDP glucose pyrophosphorylase	catalyzes the reversible production of uridine diphosphateglucose (UDPG) and diphosphate from glucose-1-phosphate (Glc-1-P) and uridine-5'-triphosphate (UTP)	R. tropici	[52]
urease	degrades urea	D. neptuniae R. lusitanum Phaseolus vulgaris-nodulating rhizobia Vigna radiata-nodulating rhizobia	[40, 46, 67]
valine acrylamidase	degrades valine-containing peptide	Bradyrhizobium spp. Rhizobium spp.	[6]
xanthine dehydrogenase	catalyzes the reaction between xanthine and NAD^+, releasing urate, NADH and H^+	R. etli R. galegae	[35, 36]

Table 1: cont….

		R. huautlense *R. tropici* *Rhizobium* spp.	
α- and β-xylosidase	degrades xyloside	*R. lusitanum*	[40]

2.3 SEROLOGICAL STUDY OF RHIZOBIA

Techniques used widely to study serological characteristics of rhizobial strains are agglutination (AG), enzyme-linked immunosorbent assay (ELISA), fluorescent-antibody (FA) test and immunodiffusion (ID). The principle of these techniques is based on the cross-reaction between somatic antigens prepared from rhizobial cells and antibody-containing antisera prepared against serotype strains. Members of a population exhibiting an identical antibody-antigen reaction are assigned to the same serogroup and can be considered similar, if not identical, organisms [68]. An antigen is any substance which induces the production of antibodies upon introduction into the blood stream of a mammal. The antigenic substances are proteins or polysaccharides, and bacteria can also serve as antigens. Three groups of antigens are recognized; somatic (cell wall) or "O", flagella or "H" and capsular or "K" antigens [69]. Antibodies are proteins produced by plasma cells (B cells) in response to antigen which the unique capability of binding specifically to the antigen induced their formation [70]. Several studies have indicated that the surface antigens of rhizobia were more specific than the flagella or internal antigens [71-73]. Sadowsky *et al.* [72] reported that there was the relationship between serological and genetic properties and symbiotic characteristics of serologically related bradyrhizobia within a given population. Rhizobial serogroups have been numbered in reference to the specific strains, especially strains deposited in the United States Department of Agriculture (USDA) collection, to which antisera were raised. Rhizobial serogroups and members are listed in Table **2**. In some cases, more than one antisera are required to characterize rhizobial strains to different serogroups. *Ensifer fredii* (formerly *Rhizobium fredii*) USDA 191, USDA 192, USDA 193, USDA 194, USDA 205 and USDA 214 could not be separated into serological groups based on their ID reactions with antisera prepared against whole-cell antigens of USDA 192, USDA

194 and USDA 205. All the strains shared at least 1 heat-labile somatic antigen. However, they could be separated into 3 distinct serogroups based on ID reactions with 3 somatic antisera [72]. Serogroups 127 and 129 shared antigenic determinants with serogroup 123 but not with each other. The term "serocluster 123" was introduced to describe the serological cross-reactivity of strains in serogroups 123, 127 and 129 [74]. Some serogroups were found predominant among field populations. *B. japonicum* serogroup 31 was predominant in soybean fields in 9 states of the United States (Arkansas, Delaware, Florida, Illinois, Iowa, Kansas, Louisiana, Minnesota and Ohio) [60, 75-79]. Serogroup 123 was also a major component of indigenous *B. japonicum* in Illinois, Iowa, Minnesota and Ohio [75-79]. *B. japonicum* serogroup 135 was predominant in Nebraska [80]. As some of *E. fredii* strains have somatic antigens in common with the genus *Bradyrhizobium* (not necessarily the same antigens), therefore this might represent an evolutionary link between these 2 divergent groups of rhizobia [72].

The serological relationships among isolates from different geographical regions have been studied. The 55 strains of *Rhizobium phaseoli* of diverse geographical origins were distinguished into 5 serogroups [81]. Sadowsky *et al.* [72] found that 11 strains of *E. fredii* isolated from geographically diverse regions of China, shared at least 1 heat-labile antigen. Sawada *et al.* [82] found that 85% of Japanese strains reacted to the antisera prepared against USDA strains and both of them, which originated in temperate regions, possessed well known cross-reaction patterns to each other. Thompson *et al.* [83] examined the serological properties of 1,500 isolates of rhizobia from tropical region in northern Thailand and showed that the response to antisera from USDA strains, which originated in temperate regions, were generally weaker than to those from Thailand. In contrast, Yokoyama *et al.* [84] reported that Thai soybean bradyrhizobial strains reacted strongly with antiserum prepared against USDA 31. The 130 indigenous *Bradyrhizobium* isolates from Thailand could be divided into 20 serogroups based on their serological reactions with antisera produced against 19 *Rhizobium* and *Bradyrhizobium* strains. Most of them (40.8%) showed no cross-reaction with any of tested antisera. It was also found that 21.5% of indigenous isolates from Thailand reacted to the antisera prepared against 5 USDA strains including USDA 31, USDA 94, USDA 110, USDA 117 and USDA 193 [85].

Table 2: Rhizobial serogroups and members within serogroups

Serogroup	Strain*	References
4	*Bradyrhizobium japonicum* USDA 4 *Bradyhizobium* spp. USDA 51 and USDA 54	[86]
6	*B. japonicum* USDA 3, USDA 6, USDA 24, USDA 44, USDA 138, USDA 147 and USDA 285 *Bradyhizobium* spp. USDA 41 and USDA 50	[86-88]
31	*Bradyrhizobium elkanii* USDA 26, USDA 29, USDA 31, USDA 33, USDA 39, USDA 40, USDA 61, USDA 67, USDA 83, USDA 116, USDA 120 and 587 *B. japonicum* USDA 270 and USDA 310	[60, 86, 88-90]
38	*B. japonicum* USDA 38 *Bradyhizobium* spp. USDA 45 and USDA 56	[86]
46	*B. elkanii* USDA 46 *B. japonicum* USDA 71a *Bradyhizobium* spp. USDA 71 and USDA 100	[86, 88]
62	*B. japonicum* USDA 62 and USDA 140	[86, 88]
76	*B. elkanii* USDA 76, USDA 117 and 29W *B. japonicum* USDA 309 *Bradyhizobium* spp. USDA 77 and USDA 103	[86, 88, 90]
94	*B. elkanii* USDA 94 and CMU28 *Bradyrhizobium* spp. USDA 93, USDA 99 and USDA 119	[85, 86, 88]
110	*B. japonicum* USDA 16, USDA 17, USDA 20, USDA 30, USDA 64, USDA 110, USDA 137, USDA 141, USDA 443, USDA 444, USDA 445, USDA 446, USDA 447, USDA 448, USDA 449, USDA 450, USDA 451, USDA 452, USDA 453, USDA 454, USDA 455, USDA 456, USDA 457, USDA 458, USDA 459, USDA 460, USDA 461, USDA 462, USDA 466, USDA 467, USDA 468, USDA 469, I-110, L1-110, WA5099-1-1 and 3I1b110	[80, 86-88, 91, 92]
117	*B. japonicum* RJ23A and 3I1b117	[80]
122	*B. japonicum* USDA 122, USDA 136 and USDA 143 *Bradyrhizobium* sp. USDA 132 *Ensifer fredii* USDA 194, USDA 217 and USDA 257	[72, 86, 88]
123	*B. japonicum* USDA 123, USDA 162, USDA 424, USDA 432, USDA 438, IA5, IA23, IA35, IN34, IN79, Lamberton 1, MN9, PRC50, OH6, OH9, OH13, St. Paul 42 and 4BAL	[74, 80, 86, 93, 94]
124	*B. japonicum* USDA 124	[86]
126	*B. japonicum* USDA 126	[86]
127	*B. japonicum* USDA 123, USDA 127, USDA 129, USDA 171, USDA 185, USDA 424, USDA 425, USDA 430, Becker 4N-18, IA44, IA67, IN56, IN64, Mn1-1c, OH5, PA3, RRC83, Rosemount 2, Waseca 1, Webster 48, WI3058 and WI3105	[74, 86, 93, 95]

Table 2: cont....

129	*B. japonicum* USDA 129, USDA 426, USDA 427, USDA 429, USDA 434, USDA 435, AK1-3a, Becker 4, Becker 5, IA3H2-6, IN78, KS6 and MS6-4a *Bradyrhizobium* sp. USDA 422	[74, 86, 93]
130	*B. elkanii* USDA 130	[86]
135	*B. japonicum* USDA 135, USDA 489, RJ10B, RJ12S, RJ17W, RJ19FY, RJ23A and 3I1b135 *Bradyrhizobium* spp. USDA 479 and USDA 490	[60, 80, 86, 88]
192	*E. fredii* USDA 192, USDA 206, USDA 217, USDA 257 and NGR234 *Ensifer meliloti* USDA 1002, USDA 1005, USDA 1027, USDA 1031, USDA 1045, USDA 1093, USDA 1098, USDA 1146 and USDA 1174 *Rhizobium* spp. Allen 770, PL146, Tal 82 and Tal 1117	[72]
194	*B. japonicum* USDA 122 and USDA 136 *E. fredii* USDA 194, USDA 201, USDA 257 and NGR234 *E. meliloti* USDA 1002, USDA 1005, USDA 1027, USDA 1031, USDA 1045, USDA 1093, USDA 1146 and USDA 1174	[72]
205	*B. elkanii* CMU16 *E. fredii* USDA 193, USDA 205, USDA 206, USDA 208, USDA 214 and NGR234 *E. meliloti* USDA 1002, USDA 1005, USDA 1027, USDA 1031, USDA 1045, USDA 1093, USDA 1146, USDA 1170 and USDA 1174	[72, 85]

*The strains USDA 31, USDA 33, USDA 46, USDA 61, USDA 76, USDA 94 and USDA 117 were formerly *B. japonicum*.
The strains USDA 16, USDA 110, RJ2S, RJ10B, RJ17W, RJ19FY, RJ23A, WA5099-1-1, 3I1b110, 3I1b117, 3I1b135 and 4BAL were formerly *Rhizobium japonicum*.
The strains USDA 192, USDA 193, USDA 194, USDA 201, USDA 205, USDA 206, USDA 208, USDA 214, USDA 217 and USDA 257 were formerly *Rhizobium fredii*.
The strain NGR234 was formerly *Rhizobium* sp. and *Rhizobium fredii*.
The strains USDA 1002, USDA 1005, USDA 1027, USDA 1031, USDA 1045, USDA 1093, USDA 1098, USDA 1146, USDA 1170 and USDA 1174 were formerly *Rhizobium meliloti*.

REFERENCES

[1] Chen WX, Yan GH, Li JL. Numerical taxonomic study of fast-growing soybean rhizobia and a proposal that *Rhizobium fredii* be assigned to *Sinorhizobium* gen. nov. Int J Syst Bacteriol 1988; 38: 392-97.

[2] Chen W, Wang E, Wang S, *et al.* Characteristics of *Rhizobium tianshanense* sp. nov., a moderately and slowly growing root nodule bacterium isolated from an arid saline environment in Xinjiang, People's Republic of China. Int J Syst Bacteriol 1995; 45: 153-59.

[3] Chen W, Tan ZY, Gao JL, *et al. Rhizobium hainanense* sp. nov., isolated from tropical legumes. Int J Syst Bacteriol 1997; 47: 870-73.

[4] Leelahawonge C, Nuntagij A, Teaumroong N, *et al.* Characterization of root-nodule bacteria isolated from the medicinal legume *Indigofera tinctoria*. Ann Microbiol 2010; 60: 65-74.

[5]   Pongsilp N, Leelahawonge C. Root-nodule symbionts of *Derris elliptica* Benth. are members of three distinct genera *Rhizobium, Sinorhizobium* and *Bradyrhizobium*. Int J Integr Biol 2010; 9: 37-42.

[6]   Pongsilp N, Leelahawonge C, Nuntagij A, *et al.* Characterization of *Pueraria mirifica*-nodulating rhizobia present in Thai soil. Afr J Microbiol Res 2010; 4: 1307-13.

[7]   Sadowsky MJ, Keyser HH, Bohlool BB. Biochemical characterization of fast- and slow-growing rhizobia that nodulate soybeans. Int J Syst Bacteriol 1983; 33: 716-22.

[8]   Xu LM, Ge C, Cui Z, *et al. Bradyrhizobium liaoningense* sp. nov., isolated from the root nodules of soybeans. Int Syst Bacteriol 1995; 45: 706-11.

[9]   Han TX, Wang ET, Wu JL, *et al. Rhizobium multihospitium* sp. nov., isolated from multiple legume species native of Xinjiang, China. Int J Syst Evol Microbiol 2008; 58: 1693-99.

[10]   Lin DX, Wang ET, Tang H, *et al. Shinella kummerowiae* sp. nov., a symbiotic bacterium isolated from root nodules of the herbal legume *Kummerowia stipulacea*. Int J Syst Evol Microbiol 2008; 58: 1409-13.

[11]   Biondi EG, Tatti E, Comparini D, *et al.* Metabolic capacity of *Sinorhizobium* (*Ensifer*) *meliloti* strains as determined by phenotype microarray analysis. Appl Envir Microbiol 2009; 75: 5396-04.

[12]   Sneath PHA, Sokal RR. Numerical Taxonomy: The Principles and Practice of Numerical Classification. W.H. Freeman and Co. San Francisco, USA. 1973.

[13]   Novikova N, Pavlova EA, Vorobjev NI, Limeshchenko EV. Numerical taxonomy of *Rhizobium* strains from legumes of the temperate zone. Int J Syst Bacteriol 1994; 44: 734-42.

[14]   Bais HP, Weir TL, Perry LG, *et al.* The role of root exudates in rhizosphere interations with plants and other organisms. Ann Rev Plant Biol 2006; 57: 233-66.

[15]   Weisskopf L, Abou-Mansour E, Fromin N, *et al.* White lupin has developed a complex strategy to limit microbial degradation of secreted citrate required for phosphate acquisition. Plant Cell Envir 2006; 29: 919-27.

[16]   Martinez-Molina E, Morales VM, Hubbell DH. Hydrolytic enzyme production by *Rhizobium.* Appl Envir Microbiol 1979; 38: 1186-88.

[17]   Chandler MR. Some observations on infection of *Arachis hypogaea* L. by *Rhizobium*. J Exp Bot 1978; 29: 749-56

[18]   Fauvart M, Verstraeten N, Dombrecht B, *et al. Rhizobium etli* HrpW is a pectin-degrading enzyme and differs from phytopathogenic homologues in enzymically crucial tryptophan and glycine residues. Microbiol 2009; 155: 3045-54.

[19]   Berthelot K, Delmotte FM. Purification and characterization of an α-glucosidase from *Rhizobium* sp. (*Robinia pseudoacacia* L.) strain USDA 4280. Appl Envir Microbiol 1999; 65: 2907-11.

[20]   Albrecht SL, Maier RJ, Hanus FJ, *et al.* Hydrogenase in *Rhizobium japonicum* increases nitrogen fixation by nodulated soybeans. Science 1979; 203: 1255-57.

[21]   Lepo JE, Hickok RE, Cantrell MA, *et al.* Revertible hydrogen uptake-deficient mutants of *Rhizobium japonicum.* J Bacteriol 1981; 146: 614-20.

[22]   Zablotowicz RM, Russell SA, Evans HJ. Effect of hydrogenase system in *Rhizobium japonicum* on the nitrogen fixation and growth of soybeans at different stages of development. Agron J 1980; 72: 555-59.

[23]   Rastogi VK, Watson RJ. Aspartate aminotransferase activity is required for aspartate catabolism and symbiotic nitrogen fixation in *Rhizobium meliloti*. J Bacteriol 1991; 173: 2879-87.

[24]   Aguilar OM, Grasso DH. The product of the *Rhizobium meliloti ilvC* gene is required for isoleucine and valine synthesis and nodulation of alfalfa. J Bacteriol 1991; 173: 7756-64.

[25]   Glenn AR, McKay IA, Arwas R, Dilworth MJ. Sugar metabolism and the symbiotic properties of carbohydrate mutants of *Rhizobium leguminosarum*. Microbiol 1984; 130: 239-45.

[26]   Ronson CW, Primrose SB. Carbohydrate metabolism in *Rhizobium trifolii*: identification and symbiotic properties of mutants. Microbiol 1979; 112: 77-88.

[27]   Drets GM, Arias A. Enzymatic basis for differentiation of *Rhizobium* into fast- and slow-growing groups. J Bacteriol 1972; 109: 467-70.

[28]   Ma W, Guinel FC, Glick BR. *Rhizobium leguminosarum* biovar viciae 1-aminocyclopropane-1-carboxylate deaminase promotes nodulation of pea plants. Appl Envir Microbiol 2003; 69: 4396-02.

[29]   Ma W, Sebestianova SB, Sebestian J, *et al.* Prevalence of 1-aminocyclopropane-1-carboxylate deaminase in *Rhizobium* spp. Antonie Leeuwenhoek 2003; 83: 285-91.

[30]   Nukui N, Minamisawa K, Ayabe SI, Aoki T. Expression of the 1-aminocyclopropane-1-carboxylic acid deaminase gene requires symbiotic nitrogen-fixing regulator gene *nifA2* in *Mesorhizobium loti* MAFF303099. Appl Envir Microbiol 2006; 72: 4964-69.

[31]   Uchiumi T, Ohwada T, Itakura I, *et al.* Expression islands clustered on the symbiosis island of the *Mesorhizobium loti* genome. J Bacteriol 2004; 186: 2439-48.

[32]   Karr DB, Waters JK, Suzuki F, Emerich DW. Enzymes of the poly-β-hydroxybutyrate and citric acid cycles of *Rhizobium japonicum* bacteroids. Plant Physiol 1984; 75: 1158-62.

[33]   Leelahawonge C, Pongsilp N. Phosphatase activities of root-nodule bacteria and nutritional factors affecting production of phosphatases by representative bacteria from three different genera. KMITL Sci Technol J 2009; 9: 65-83.

[34]   Pantujit S, Pongsilp N. Phosphatase activity and effects of phosphate-solubilizing bacteria on yield and uptake of phosphorus in corn. World Appl Sci J 2010; 8: 429-35.

[35]   Diouf A, de Lajudie P, Neyra M, *et al.* Polyphasic characterization of rhizobia that nodulate *Phaseolus vulgaris* in West Africa (Senegal and Gambia). Int J Syst Evol Microbiol 2000; 50: 159-70.

[36]   Wang ET, van Berkum P, Beyene D, *et al. Rhizobium huautlense* sp. nov., a symbiont of *Sesbania herbacea* that has a close phylogenetic relationship with *Rhizobium galegae*. Int J Syst Bacteriol 1998; 48: 687-99.

[37]   Irigoyen JJ, Sanchez-Diaz M, Emerich DW. Carbon metabolism enzymes of *Rhizobium meliloti* cultures and bacteroids and their distribution within alfalfa nodules. Appl Envir Microbiol 1990; 56: 2587-89.

[38]   Leong SA, Ditta GS, Helinski DR. Heme biosynthesis in *Rhizobium*. Identification of a cloned gene coding for delta-aminolevulinic acid synthetase from *Rhizobium meliloti*. J Biol Chem 1982; 257: 8724-30.

[39]   Watson RJ, Rastogi VK. Cloning and nucleotide sequencing of *Rhizobium meliloti* aminotransferase genes: an aspartate aminotransferase required for symbiotic nitrogen fixation is atypical. J Bacteriol 1993; 175: 1919-28.

[40]   Valverde A, Igual JM, Peix A, *et al. Rhizobium lusitanum* sp. nov. a bacterium that nodulates *Phaseolus vulgaris*. Int J Syst Evol Microbiol 2006; 56: 2631-37.

[41]   Huerta-Zepeda A, Duran S, Du Pont G, Calderon J. Asparagine degradation in *Rhizobium etli*. Microbiol 1996; 142: 1071-76.

[42]   Poole PS, Dilworth MJ, Glenn AR. Acquisition of aspartase activity in *Rhizobium leguminosarum* WU235. Microbiol 1984; 130: 881-86.

[43]   Parke D, Ornston LN. Enzymes of the beta-ketoadipate pathway are inducible in *Rhizobium* and *Agrobacterium* spp. and constitutive in *Bradyrhizobium* spp. J Bacteriol 1986; 165: 288-92.

[44]   Mateos PF, Jimenez-Zurdo JI, Chen J, *et al.* Cell-associated pectinolytic and cellulolytic enzymes in *Rhizobium leguminosarum* biovar trifolii. Appl Envir Microbiol 1992; 58: 1816-22.

[45]   Sudto A, Punyathiti Y, Pongsilp N. The use of agricultural wastes as substrates for cell growth and carboxymethyl cellulase (CMCase) production by *Bacillus subtilis*, *Escherichia coli* and *Rhizobium* sp. KMITL Sci Tech J 2008; 8: 84-92.

[46]   Rivas R, Willems A, Subba-Rao NS, *et al.* Description of *Devosia neptuniae* sp. nov. that nodulates and fixes nitrogen in symbiosis with *Neptunia natans*, an aquatic legume from India. Syst Appl Microbiol 2003; 26: 47-53.

[47]   Yao ZY, Kan FL, Wang ET, *et al.* Characterization of rhizobia that nodulate legume species of the genus *Lespedeza* and description of *Bradyrhizobium yuanmingense* sp. nov. Int J Syst Evol Microbiol 2002; 52: 2219-30.

[48]   Robledo M, Jimenez-Zurdo JI, Velazquez E, *et al.* *Rhizobium* cellulase CelC2 is essential for primary symbiotic infection of legume host roots. Proc Natl Acad Sci USA. 2008; 105: 7064-69.

[49]   Wang, ET, van Berkum P, Sui XH, *et al.* Diversity of rhizobia associated with *Amorpha fruticosa* isolated from Chinese soils and description of *Mesorhizobium amorphae* sp. nov. Int J Syst Bacteriol 1999; 49: 51-65.

[50]   Roy N, Bhattacharyya P, Chakrabartty PK. Iron acquisition during growth in an iron-deficient medium by *Rhizobium* sp. isolated from *Cicer arietinum*. Microbiol 1994; 140: 2811-20.

[51]   Fennington GJ Jr, Hughes TA. The fructokinase from *Rhizobium leguminosarum* biovar *trifolii* belongs to group I fructokinase enzymes and is encoded separately from other carbohydrate metabolism enzymes. Microbiol 1996; 142: 321-30.

[52]   Romanov VI, Hernandez-Lucase E, Martinez-Romerro E. Carbon metabolism enzymes of *Rhizobium tropici* cultures and bacteroids. Appl Envir Microbiol 1994; 60: 2339-42.

[53]   Abe M, Higashi S. β-glucosidase and β-galactosidase from the periplasmic space of *Rhizobium trifolii* cells. J Gen Appl Microbiol 1982; 28: 551-62.

[54]   Charles TC, Singh RS, Finan TM. Lactose utilization and enzymes encoded by megaplasmids in *Rhizobium meliloti* SU47: implications for population studies. J Gen Microbiol 1990; 136: 2497-02.

[55]   Zurdo-Pineiro JL, Rivas R, Trujillo ME, *et al. Ochrobactrum cytisi* sp. nov., isolated from nodules of *Cytisus scoparius* in Spain. Int J Syst Evol Microbiol 2007; 57: 784-88.

[56]   Duran S, Calderon J. Role of the glutamine transaminase-omega-amidase pathway and glutaminase in glutamine degradation in *Rhizobium etli*. Microbiol 1995; 141: 589-95.

[57]   Zorreguieta A, Finnie C, Downie JA. Extracellular glycanases of *Rhizobium leguminosarum* are activated on the cell surface by an exopolysaccharide-related component. J Bacteriol 2000; 182: 1304-12.

[58]   Morales V, Martinez-Molina E, Hubbell D. Cellulase production by *Rhizobium*. Plant Soil 1984; 80: 407-15.

[59]   Harker AR, Xu LS, Hanus FJ, Evans HJ. Some properties of the nickel-containing hydrogenase of chemolithotrophically grown *Rhizobium japonicum*. J Bacteriol 1984; 159: 850-56.

[60]   Keyser HH, Weber DF, Uratsu SL. *Rhizobium japonicum* serogroup and hydrogenase phenotype distribution in 12 states. Appl Envir Microbiol 1984; 47: 613-15.

[61]   Wong CM, Dilworth MJ, Glenn AR. Cloning and sequencing show that 4-hydroxybenzoate hydroxylase (PobA) is required for uptake of 4-hydroxybenzoate in *Rhizobium leguminosarum*. Microbiol 1994; 140: 2775-86.

[62]   Driscoll BT, Finan TM. Properties of NAD^+- and $NADP^+$-dependent malic enzymes of *Rhizobium* (*Sinorhizobium*) *meliloti* and differential expression of their genes in nitrogen-fixing bacteroids. Microbiol 1997; 143: 489-98.

[63]   Prasuna P, Ali S. Detection and characterization of two thermally reactive pectinases in cultures of *Rhizobium*. Indian J Exp Biol 1987; 25: 632-33.

[64]   Basu SS, York JD, Raetz CRH. A phosphotransferase that generates phosphatidylinositol 4-phosphate (PtdIns-4-P) from phosphatidylinositol and lipid A in *Rhizobium leguminosarum*: a membrane-bound enzyme linking lipid A and PtdIns-4-P biosynthesis. J Biol Chem 1999; 274: 11139-49.

[65]   Munoz JA, Coronado C, Perez-Hormaeche J, *et al*. MsPG3, a *Medicago sativa* polygalacturonase gene expressed during the alfalfa–*Rhizobium meliloti* interaction. Proc Natl Acad Sci USA. 1998; 95: 9687-92.

[66]   Magne O, Driscoll BT, Finan TM. Increased pyruvate orthophosphate dikinase activity results in an alternative gluconeogenic pathway in *Rhizobium* (*Sinorhizobium*) *meliloti*. Microbiol 1997; 143: 1639-48.

[67]   Chuntanom S, Pongsilp N. Environmental parameters affecting urease production and ammonification in *Phaseolus vulgaris*-nodulating and *Vigna radiata*-nodulating rhizobia. Int J Microbiol Res 2011; 2: 222-32.

[68]   Brockman FJ, Bezdicek DF. Diversity within serogroups of *Rhizobium leguminosarum* biovar *viceae* in the Palouse region of Eastern Washington as indicated by plasmid profiles, intrinsic antibiotic resistance, and topography. Appl Envir Microbiol 1989; 55: 109-15.

[69]   Somasegaran P, Hoben HJ. Handbook for Rhizobia: Methods in Legume-*Rhizobium* Technology. NIFTAL Project, University of Hawaii, Paia, USA. 1994.

[70]   Subba Rao NS. Soil Microbiology (Fourth Edition of Soil Microorganisms and Plant Growth). Science Publishers Inc., New Hamspire, USA. 1999.

[71]   Pankhurst CE. Some antigenic properties of cultured cells and bacteroid forms of fast- and slow-growing strains of *Lotus* rhizobia. Microbios 1979; 24: 19-28.

[72]   Sadowsky MJ, Bohlool BB, Keyser HH. Serological relatedness of *Rhizobium fredii* to other rhizobia and to the bradyrhizobia. Appl Envir Microbiol 1987; 53: 1785-89.

[73]   Vincent JM. Serology. In: Broughton WJ, Ed. Nitrogen Fixation, Vol 2. Oxford, Clarendon Press, 1982; pp. 235-73.

[74]   Schmidt EL, Zidwick MJ, Abebe HM. *Bradyrhizobium japonicum* serocluster 123 and diversity among member isolates. Appl Envir Microbiol 1986; 51: 1212-15.

[75]   Damirgi SM, Frederick LR, Anderson IC. Serogroups of *Rhizobium japonicum* in soybean nodules as affected by soil types. Agron J 1967; 59: 10-12.

[76]   Ham GE, Frederick LR, Anderson IC. Serogroups of *Rhizobium japonicum* in soybean nodules sampled in Iowa. Agron J 1971; 63: 69-72.

[77]   Kapusta G, Rouwenhorst DL. Influence of inoculums size on *Rhizobium japonicum* serogroup distribution frequency in soybean nodules. Agron J 1973; 65: 916-19.

[78]   Kvien CS, Ham GE, Lambert JW. Recovery of introduced *Rhizobium japonicum* strains by soybean genotypes. Agron J 1981; 73: 900-05.

[79]   Reyes VG, Schmidt EL. Population densities of *Rhizobium japonicum* strain 123 estimated directly in soil and rhizospheres. Appl Envir Microbiol 1979; 37: 854-58.

[80]   Gross DC, Vidaver AK, Klucas RV. Plasmids, biological properties and efficacy of nitrogen fixation in *Rhizobium japonicum* strains indigenous to alkaline soils. Microbiol 1979; 114: 257-66.

[81]   Robert FM, Schmidt EL. Somatic serogroups among 55 strains of *Rhizobium phaseoli*. Can J Microbiol 1985; 31:519-23.

[82]   Sawada Y, Miyashita K, Tanabe I, Kato K. *Hup* phenotype and serogroup identity of soybean-nodulating bacteria isolated from Japanese soil. Soil Sci Plant Nutr 1989; 35: 281-88.

[83]   Thompson JA, Bhromsiri A, Shutsrirung A, Lillakan S. Native root-nodule bacteria of traditional soybean-growing areas of northern Thailand. Plant Soil 1991; 135: 53-65.

[84]   Yokoyama T, Ando S, Murakami T, Imai H. Genetic variability of the common *nod* gene in soybean bradyrhizobia isolated in Thailand and Japan. Can J Microbiol 1996; 42: 1209-18.

[85]   Pongsilp N, Teaumroong N, Nuntagij A, *et al.* Genetic structure of indigenous non-nodulating and nodulating populations of *Bradyrhizobium* in soils from Thailand. Symbiosis 2002; 33: 39-58.

[86]   van Berkum P, Fuhrmann JJ. Evidence from internally transcribed spacer sequence analysis of soybean strains that extant *Bradyrhizobium* spp. are likely the products of reticulate evolutionary events. Appl Envir Microbiol 2009; 75: 78-82.

[87]   Date RA, Decker AM. Minimal antigenic constitution of 28 strains of *Rhizobium japonicum*. Can J Microbiol 1965; 11: 1-8.

[88]   van Berkum P, Keyser HH. Anaerobic growth and denitrification among different serogroups of soybean rhizobia. Appl Envir Microbiol 1985; 49: 772-77.

[89]   Fuhrmann J, Wollum II AG. Simplified enzyme linked immunosorbent assay for routine identification of *Rhizobium japonicum* antigens. Appl Envir Microbiol 1985; 49: 1010-13.

[90]   Rumjanek NG, Dobert RC, van Berkum P, Triplett EW. Common soybean inoculant strains in Brazil are members of *Bradyrhizobium elkanii*. Appl Envir Microbiol 1993; 59: 4371-73.

[91]   Basit HA, Angle JS, Salem S, *et al.* Phenotypic diversity among strains of *Bradyrhizobium japonicum* belonging to serogroup 110. Appl Envir Microbiol 1991; 57: 1570-72.

[92]   van Berkum P, Kotob SI, Basit HA, *et al.* Genotypic diversity among strains of *Bradyrhizobium japonicum* belonging to serogroup 110. Appl Envir Microbiol 1993; 59: 3130-33.

[93]   Judd AK, Schneider M, Sadowsky MJ, de Bruijn FJ. Use of repetitive sequences and the polymerase chain reaction technique to classify genetically related *Bradyrhizobium japonicum* serocluster 123 strains. Appl Envir Microbiol 1993; 59: 1702-08.

[94]   Rodriguez-Quinones F, Judd AK, Sadowsky MJ, *et al.* Hyperreiterated DNA regions are conserved among *Bradyrhizobium japonicum* serocluster 123 strains. Appl Envir Microbiol 1992; 58: 1878-85.

[95]   Sadowsky MJ, Cregan PB, Keyser HH. DNA hybridization probe for use in determining restricted nodulation among *Bradyrhizobium japonicum* serocluster 123 field isolates. Appl Envir Microbiol 1990; 56: 1768-74.

CHAPTER 3

Symbiotic Variation and Plant-Growth-Promoting Traits of Rhizobia

Neelawan Pongsilp[*]

Department of Microbiology, Faculty of Science, Silpakorn University, Nakhon Pathom, Thailand

Abstract: The rhizobia-legume symbioses exhibit variation in symbiotic performance as measured by plant yield, nodulation and nitrogenase activity. The previous studies have demonstrated that variation in symbiotic performance is dependent on both rhizobial strains and plant species (or cultivars). It has also been found that symbiotic variation is connected to some characteristics of rhizobia including serological and morphological phenotypes, tolerance to stresses, host range, plasmid profile as well as some cryptic plasmids. Rhizobia are important members of plant-growth-promoting rhizobacteria (PGPR) that exert the positive effects on plant growth *via* direct and indirect mechanisms. Plant-growth-promoting (PGP) activities include the production of phytohormones, siderophores and 1-aminocyclopropane-1-carboxylic acid (ACC) deaminase as well as the solubilization of inorganic phosphate. Many rhizobial strains produce phytohormones such as indole-3-acetic acid (IAA) with auxin activity and gibberellins (GAs). The role of IAA produced by rhizobia on plant-growth promotion has been demonstrated. GAs are involved in the formation of infection pocket and infection thread. Siderophores are iron chelators that chemically bind and solubilize iron. The production of siderophores is connected to the decrease or prevention of deleterious effect of pathogenic microorganisms. Phosphate-solubilizing activity of rhizobia can increase phosphorus availability to plants. The plant-growth promotion by rhizobia with phosphate-solubilizing activity has been reported. The ACC-producing rhizobia promote nodulation and root elongation of their host plants. These PGP activities make rhizobia superior PGPR for legumes and non-legumes.

Keywords: 1-Aminocyclopropane-1-carboxylic acid (ACC) deaminase, Gibberellin (GA), Host range, Indole-3-acetic acid (IAA), Phosphate solubilization, Plant-growth-promoting rhizobacteria (PGPR), Siderophore, Symbiotic effectiveness, Symbiotic variation, Rhizobia.

3.1. SYMBIOTIC VARIATION OF RHIZOBIA

Up to the present time, many rhizobial strains have been found to exhibit a broad host range. *Ensifer fredii* (formerly *Rhizobium* sp.) NGR234 possessed the boardest

---

*Address correspondence to Neelawan Pongsilp:** Department of Microbiology, Faculty of Science, Silpakorn University, Nakhon Pathom, Thailand; Tel: +66-34-245337; Fax: +66-34-245336-37; E-mail: neelawan@su.ac.th

host range which includes 232 legume species [1]. In pararell, many host plants can be nodulated by rhizobia which belong to different taxonomic groups. *Phaseolus vulgaris* (common bean) can be nodulated by rhizobia in at least 33 species (distributed in 11 genera) including *Azorhizobium caulinodans*, *Azorhizobium doebereinerae*, *Bradyrhizobium japonicum*, *Bradyrhizobium liaoningense*, *Burkholderia phymatum*, *Cupriavidus* spp./ *Ralstonia* spp., *E. fredii*, *Ensifer medicae*, *Ensifer meliloti*, *Ensifer* spp./*Sinorhizobium* spp., *Herbaspirillum lusitanum*, *Mesorhizobium caraganae*, *Mesorhiobium septentrionale*, *Mesorhizobium shangrilense*, *Mesorhizobium temperatum*, *Methylobacterium* spp., *Rhizobium alkalisoli*, *Rhizobium etli*, *Rhizobium gallicum*, *Rhizobium giardinii*, *Rhizobium huautlense*, *Rhizobium leguminosarum*, *Rhizobium leucaenae*, *Rhizobium lusitanum*, *Rhizobium miluonense*, *Rhizobium phaseoli*, *Rhizobium tibeticum*, *Rhizobium tropici*, *Rhizobium vallis*, *Rhizobium yanglingense* and *Rhizobium* spp. *Macroptilium atropupureum* (siratro) is a promiscuous host for rhizobia in at least 18 species (distributed in 9 genera) including *Azo. doebereinerae*, *Bradyrhizobium canariense*, *Bradyrhizobium elkanii*, *Bradyrhizobium japonicum*, *Burkholderia tuberum*, *Cupriavidus taiwanensis* (formerly *Ralstonia taiwanensis)*, *Cupriavidus* spp./*Ralstonia* spp., *E. fredii*, *Ensifer* spp./*Sinorhizobium* spp., *Mesorhizobium australicum*, *Mesorhizobium opportunistum*, *Mes. septentrionale*, *Rhi. etli*, *Rhi. gallicum*, *Rhi. tropici* and *Rhizobium* spp. Among the associated partners, variation in symbiotic performance has been observed. The parameters for measurement of symbiotic performance include plant yield, nodulation and nitrogenase activity [2-5]. The symbiosis between rhizobia and their hosts is a consequence, in which genetic differences in the ability to develop and sustain a significant relationship may occur at 2 levels, in the host and in the bacterium [6]. Thus, variation in symbiotic performance is dependent on both rhizobial strains and plant species (or cultivars). For example, *E. fredii* (formerly *Rhizobium fredii*) USDA 257 did not nodulate legume species such as *Dalea candida* Willd., *Glycine canescens* Herman, *Indigofera jamaicensis* Spreng., *Leucaena leucocephala* (Lam.) DeWit (white popinac) and *Phaseolus polystachyus* Britt. Stearns & Pogg, but *E. fredii* NGR234 effectively nodulated these legume hosts. As compared with 232 legume hosts of NGR234, USDA 257 effectively and ineffectively nodulated 135 legume species. The host range of USDA 257 is a subset nested entirely within that of NGR234.

Both strains, NGR234 and USDA 257, did not nodulate legume cultivars such as *Phaseolus coccineus* L. (scarlet runner bean), *Pha. vulgaris* L. and *Vigna radiata* (mungbean) subsp. *sublobata,* but effectively nodulated the other cultivars *Pha. coccineus* L. subsp. *polyanthus* (Greenman) Marechal, *Pha. vulgaris* L. var. *aborigineus* (Burk.) Baudet and *V. radiata* (L.) Wilczek (Roxb.) Verdc. Mascherpa, & Stainier [1].

It has been found that the variation in symbiotic effective results from the combination of rhizobial isolates and plant species (or cultivars). For example, 10 strains of *Bradyrhizobium japonicum* varied in their biological and symbiotic properties with *Glycine max* (soybean) cultivar Amsoy 71, with values of acetylene reduction activity (ARA) ranging from 55.5 to 162.4 nmol ethylene/plant/min [7]. The 4 cultivars of *Medicago sativa* (alfalfa) exhibited significantly different symbiotic responses to 5 *E. meliloti* isolates [8]. *E. fredii* NGR234 formed ineffective nodules on *G. max* cultivar Caloria, but it failed to nodulate cultivar Peking and cultivar McCall [9, 10]. This result suggests that *G. max* is a good example of the kinds of problems presented by suboptimal growth conditions. Somasegaran and Martin [11] examined symbiotic characteristics of the leucaenas (*L. leucocephala*, *Leucaena diversitolia* and the hybrid of *L. leucocephala* and *L. diversifolia*). Among 3 strains of *Rhizobium* sp. (TAL 582 TAL 820 and TAL 1145), the most effective and competitive *Rhizobium* sp. for the leucaenas was TAL 1145. *L. leucocephala* and the hybrid showed 100% more total nitrogen than did *L. diversifolia*. The symbiotic characters of the hybrid closely resembled to those of *L. leucocephala* more than those of *L. diversifolia*. Symbiotic effectiveness of the fast-growing *E. fredii* USDA 191 was compared with that of the slow-growing *Bradyrhizobium japonicum* USDA 110 in symbioses with 5 cultivars of *G. max* including 4 North American cultivars and 1 Chinese cultivar. The strain USDA 110 fixed 3.7 to 57.3 times more nitrogen than did USDA 191with all North American cultivars. The strain USDA 191 fixed 3.3 times more nitrogen than did USDA 110 with 1 Chinese cultivar. The superior nitrogen fixation capability of USDA 110 with 4 North American cultivars resulted primarily from higher nitrogenase activity per nodule mass (specific acetylene reduction activity) and higher nodule mass per plant [12]. Numerous *Rhizobium trifolii* strains exhibited varying levels of symbiotic effectiveness on 5 species of African annual *Trifolium* (clover) including *T. decorum*, *T. quartinianum*, *T. rueppellianum*, *T. steudneri* and *T. tembense* [13]. Significant variation in symbiotic performance was found among 22 *Acacia* species inoculated with different rhizobial isolates. Rhizobial isolates were shown to be highly, moderately or weakly successful in forming an effective association. For most species, the mean effectiveness of any isolates was 15% to 20% lower than that of the most effectiveness combination [6]. *Vigna unguiculata* (cowpea) varied greatly with estimated values of effectiveness, ranging from 23.5% to 118.0% when compared with the uninoculated controls [4]. Among 34 isolates of *Rhi. etli* and *Rhi. gallicum* that nodulated and fixed nitrogen with *Pha. vulgaris*, *Rhi. gallicum* isolates induced only one-third of the nodules induced by *Rhi. etli* isolates. All *Rhi. gallicum* isolates formed effective nodules on *Mac. atropurpureum*, whereas *Rhi. etli* isolates induced less than one-third of the nodules induced by *Rhi. gallicum* isolates on this host and were ineffective [14]. Jones [5] observed symbiotic variation of *Rhi. trifolii* isolates from mountain soils and lowland soils on *Trifolium repens* L. (white clover). Appunu *et al.* [2] reported the variable extent of nitrogen fixation by *Bradyrhizobium japonicum* strains and *G. max* cultivars that is probably due to differences in symbiotic effectiveness of rhizobial strains and their compatibility.

It has been found that symbiotic variation is connected with some phenotypic and genotypic characteristics. Symbiotic effectiveness was found to be related to serological and morphological phenotypes of rhizobia. Among 360 isolates of *Bradyrhizobium japonicum* that nodulate *G. max*, the nitrogen content of plant shoots was strongly and comparably related to both serological and morphological groupings, while rhizobitoxine and hydrogenase phenotypes were relatively poor predictors of symbiotic effectiveness [15]. Bradyrhizobial strains that were more tolerant to stress were more effective on symbiotic nitrogen fixation with *V. unguiculata* under acid-soil conditions [3]. All *Bradyrhizobium japonicum* strains identical in plasmid number and size also had approximately the same rate of ethylene production in symbiosis with *G. max* cultivar Amsoy 71 [7].

There are some hypotheses in connection with the variation in symbiotic effectiveness. One explanation is a trade-off between symbiotic effectiveness and host range of rhizobia. Burdon *et al.* [6] hypothesized that wider host ranges of rhizobia are linked with lower levels of effectiveness on any one host. The strain of *Rhi. trifolii* able to fix nitrogen in root nodules of *Trifolium pratense* (red clover) lost its effectiveness when treated with DNA from an ineffective strain. The result supports the idea that symbiotic effectiveness involves compatability between several plant and bacterial factors, changes in any one of which makes the bacterium ineffective [16]. Significant variation in the nitrogen-fixation capabilities of 4 North American cultivars of *G. max* inoculated with *E. fredii* USDA 191 was also observed. This quantitative variation in nitrogen-fixation capability suggests that the total incompatibility (effectiveness of nodulation and efficiency of nitrogen-fixation) of the host cultivars and *E. fredii* strains is regulated by more than one host plant gene [12]. In some rhizobia such as *E. meliloti* (formerly *Rhizobium meliloti*), genes coding for nitrogen fixation and nodulation are located on the symbiotic megaplasmids (pSym). Besides pSym, the cryptic plasmids encoding genes for other functions are also present. The relation between the symbiotic effectiveness and the presence of cryptic plasmid was found in the previous studies. One of the cryptic plasmids (pRmeGR4b) had a positive effect on nodulation capacity of *E. meliloti* GR4 [17]. *E. meliloti* SAF22 had the genetic capability to develop fully effective root nodules on *Med. sativa* cultivar Aragon but this phenotype was attenuated by its cryptic plasmid, pRmSAF22c, which interfered with the normal nodular development necessary to sustain a fully effective nitrogen-fixing symbiosis. Due to the presence of pRmeSAF22c, this strain was not fully genetically compatible with *Med. sativa* and the elimination of this cryptic plasmid improved the symbiotic performance as measured by the increase in lengths, dry weights and nitrogen contents of inoculated plants [18].

3.2. PLANT-GROWTH-PROMOTING (PGP) TRAITS OF RHIZOBIA

Rhizobia are important members of plant-growth-promoting rhizobacteria (PGPR) showing several plant-growth-promoting (PGP) activities. PGPR can exert the positive effects on plant growth *via* direct and indirect mechanisms. Direct growth promotion results from the nitrogen fixation, the production of phytohormones and some enzymes such as 1-aminocyclopropane-1-carboxylic acid (ACC) deaminase that modulates the level of phytohormones as well as the solubilization of inorganic phosphate which makes phosphorous available to the plants. Indirect growth promotion results from the decrease or prevention of deleterious effect of pathogenic microorganisms, mostly due to the production of siderophores or antibiotics.

3.2.1. Production of Phytohormones by Rhizobia

Phytohormones control all developmental plant processes including the nodulation that is presumably initiated by a change in the cytokinin:auxin ratio within the root [19]. In rhizobia-legume symbiosis, it has been found that the fixed nitrogen is not the only benefit a host plant receives from rhizobia. The rhizobial partner may provide growth-regulating factors such as phytohormones [20].

3.2.1.1. Production of Indole-3-Acetic Acid (IAA) by Rhizobia

Indole-3-acetic acid (IAA) is the main plant growth hormone with auxin activity [21]. IAA plays a key role in the control of many physiological processes in plants such as root proliferation [22-24], cell division and shoot growth [21]. IAA production is known to be involved in processes of symbiotic bacteria [25, 26], plant-associated bacteria [27, 28] and pathogenic bacteria [21, 27, 29]. The production of IAA, a phytohormone that does not apparently function as a hormone in bacterial cells, may have evolved in bacteria because it is important in the bacterium-plant relationship [24]. IAA alone or in conjunction with other phytohormones might be involved in several stages of the symbiotic relationship [30, 31]. Nod factor-induced inhibition of auxin transport would lead to the local accumulation of auxins needed to trigger a nodule primordium [32].

IAA biosynthesis in rhizobia has been reported to occur through 3 pathways that have been characterized at the molecular level: 1) the indole-3-pyruvic acid (IPA) pathway [tryptophan → indole-3-pyruvic acid (IPA)→ indole-3-acetaldehyde → IAA], in which IPA intermediate is derived from tryptophan by an aminotransferase activity [33]; 2) the indole-3-acetamide (IAM) pathway [tryptophan → indole-3-acetamide (IAM) → IAA], in which tryptophan 2-monooxygenase catalyzes the conversion of tryptophan to IAM, then IAM hydrolase catalyzes the conversion of IAM to IAA [34]; 3) the indole-3-acetonitrile (IAN) pathway [tryptophan → indole-3-acetaldoxime (IAOx) →

indole-3-acetonitrile (IAN) → IAA], in which nitrile hydratase and amidase catalyzes the conversion of IAN to IAA [35, 36]. The IPA pathway was found in *E. meliloti* [33] and *Bradyrhizobium elkanii* [37]. The IAM pathway was found in *Bradyrhizobium* spp. [38] and *Rhi. leguminosarum* [39]. The IAN pathway was found in *E. meliloti*, *Mesorhizobium loti* (formerly *Rhizobium loti*) and *Rhi. leguminosarum* [36].

Rhizobial isolates have the capacity to produce IAA in culture supplemented with an amino acid, tryptophan. These isolates have been obtained from various leguminous plants including *Acacia mangium* [40], *Alysicarpus vaginalis* [41], *Cajanus cajan* (pigeon pea) [42], *Derris elliptica* [26], *Desmanthus virgatus* (wild tantan) [40], *Desmodium gangeticum* (sarivan) [43], *G. max* [40], *Indigofera tinctoria* (true indigo) [26], *Mucuna pruriens* [44], *Neptunia* sp. [40], *Phaseolus mungo* (*Vigna mungo*; black gram) [45], *Pterocarpus macrocapus* [40], *Pueraria mirifica* (white Kwao Kruea) [26], *Sesbania aculeata* (*Sesbania bispinosa*; prickly sesban), *Sesbania rostrata*, *Sesbania speciosa* and *V. radiata* [40]. The production of IAA by rhizobia has been found to vary greatly among different species and strains. In all isolates tested, the growth and the IAA production started simultaneously and the highest level of IAA was produced in the stationary phase of growth [40, 46]. The 50 rhizobial isolates from 10 leguminous plants produced IAA varying from 0.60 ± 0.14 to 25.16 ± 0.81 µg/ml [40]. The 56 rhizobial strains isolated from 3 medicinal legumes including *Der. elliptica, I. tinctoria* and *Pue. mirifica* varied in their ability to produce IAA ranging from 4.31 ± 0.67 to 34.76 ± 0.18 µg/ml [26].

The role of IAA produced by rhizobia on plant-growth promotion has been demonstrated. Besides their legume hosts, rhizobia are also superior colonizers for non-legumes. *Rhi. leguminosarum* has been reported for root colonization and growth promotion of several non-legumes such as lettuce [47, 48], maize [47-51], rape [50, 51], rice [52], sugar beet [50, 51] and wheat [47, 48, 50, 51]. *Sinorhizobium* sp. AS014 isolated from agricultural soil stimulated the growth of *Brassica oleracea* and *Raphanus sativus* (radish) with averages 5.63-fold and 5.33-fold over controls for root yields of *Brassica oleracea* and *Rap. sativus*, respectively, also increased up to 2.95-fold and 3.34-fold over controls for shoot yields of *Brassica oleracea* and *Rap. sativus*, respectively. These values did not differ significantly from the treatments applied with IAA 50 μg/ml, suggesting that the enhancement of the plant growth upon the inoculation of the IAA-producing strains of rhizobia has been attributed to the production of IAA [53].

The role of IAA produced by rhizobia on nodule formation and nitrogen fixation has been investigated. Some studies have reported the positive effects of IAA on the symbiotic processes. The increased symbiotic nitrogen fixation has been reported for high-IAA-producing mutants of *Bradyrhizobium japonicum* [54] and *E. meliloti* (formerly *Sinorhizobium meliloti*) 1021 [55]. The high-IAA-producing mutants of *E. meliloti* 1021 also showed higher levels of acid phosphatase and organic acid excretion in mobilizing phosphate from insoluble sources such as phosphate rock [55]. *E. meliloti* Rm1021 mutants with altered response to tryptophan analogs still produced IAA, but could not fix nitrogen because the mutants did not fully differentiate into the nitrogen-fixing bacteroid form [46]. In contrast, similar IAA concentrations were found in culture supernatants of *Rhizobium* mutants which lack nodulation genes (defective in nodule formation) and their corresponding parent strains, suggesting that root hair curling is not simply a consequence of IAA production by rhizobia [56]. The high-IAA-producing mutants of *Bradyrhizobium japonicum* were poor symbiotic nitrogen fixers. Plants inoculated with the mutants had a lower nodule mass and fixed less nitrogen per gram of nodule than did plants inoculated with the wild-type strain [57]. Mutants of *E. meliloti* 102F34 which lack aminotransferase activity, a key enzyme in IAA biosynthesis, were symbiotically equivalent to the wild-type strain F34 [33]. The influence of IAA produced by rhizobia on stress tolerance has been reported. The IAA-overproducing mutant strain of *E. meliloti* 1021 accumulated a higher level of trehalose as its endogenous osmolyte and showed an increased tolerance to several stress conditions (55°C, 4°C, pH 3.0, 0.5 M NaCl and UV-irradiation). *Medicago truncatula* improved salt tolerance when nodulated by the IAA-overproducing strain of *E. meliloti* [58].

3.2.1.2. Production of Gibberellins (GAs) by Rhizobia

Gibberellins (GAs) comprise a large group of more than 130 diterpenoid carboxylic acids, of which most are precursors or inactivated forms, but some members, including GA_1, GA_3, GA_4 and GA_7, have an intrinsic growth-promoting activity [19]. These GAs are involved in several processes including seed germination, seedling emergence, stem and leaf growth, floral induction as well as flower and fruit growth [59]. GAs are also involved in the induction of several genes necessary for the synthesis and secretion of α-amylase before germination of seeds [60]. The biochemistry of GA biosynthesis can be subdivided into 3 main stages. In the first stage, geranylgeranyl diphosphate is converted by 2 terpene cyclases to *ent*-kaurene, which, in the next stage, is oxidized by cytochrome P450 monooxygenases to yield GA_{12} and GA_{53} that, in the final stage, are converted to bioactive GAs by 2-oxoglutarate-dependent dioxygenases, GA 20-oxidase (GA20ox) and GA 3β-hydroxylases (GA3ox) [61, 62].

Rhizobia such as *E. meliloti* and *Rhi. phaseoli* have the capacity to produce GA-like substances [63, 64]. The levels of GA-like substances are generally higher in nodules than in nearby root tissue [64]. The *SrGA20ox1* gene encoding GA20ox, a key enzyme in GA biosynthesis, was transiently up-regulated during lateral root base nodulation of *Ses. rostrata* infected by *Azo. caulinodans*. It has been demonstrated that GAs are Nod factor downstream signals for nodulation in the hydroponic growth as well as involved in the formation of infection pocket and infection thread [19].

3.2.2. Production of Siderophores by Rhizobia

Iron is an essential nutrient for the growth and proliferation of bacteria [65]. Although relatively abundant, in its normal ferric ion (Fe^{3+}) form, it is very insoluble at normal pH [66]. When grown in iron-deficient conditions, many bacteria synthesize and release iron chelators, termed as "siderophores", that are chemically capable of binding and solubilizing iron [67]. The rhizobia-legume symbiosis is iron dependent as iron is a structural component of key proteins such as nitrogenase, hydrogenase, ferredoxin (a protein that mediates electron transfer and provides energy for nitrogen fixation) and leghaemoglobin (an iron-rich protein that buffers the oxygen level to protect the oxygen-labile nitrogenase) [68, 69]. It has been hypothesized that siderophores in rhizobia play a role in the competition in the rhizosphere, perhaps in a manner similar to that of siderophores in *Pseudomonas* strains [70]. Rhizobial species produce various types of siderophores, including hydroxamate, carboxylate and catecholate, as shown in Table **1**.

The role of siderophore production on the nodule formation and the nitrogen fixation has been argumentative. The biosynthesis of anthranilate, that acts as an *in planta* siderophore, was found to be necessary during nodule development for the establishment of an effective symbiosis between *E. meliloti* and *Med. sativa*. Mutants unable to synthesize anthranilate displayed 2 symbiotic phenotypes. One type contained bacteroids and was capable of nitrogen fixation, while the other lacked bacteroids and could not fix nitrogen [71]. In contrast, genes that function in the biosynthesis and the transport of rhizobactin 1021, a siderophore produced by *E. meliloti*, were not strongly expressed when nitrogenase was being formed in root nodules. Mutants having transposon insertions in genes involved in the biosynthesis or the transport still induced effective nodules on its host, *Med. sativa* [69]. The *fur* (ferric uptake regulator) mutant of *E. meliloti*, that is defective in hydrogen peroxide sensitivity, manganese resistance and siderophore over-production, nodulated *Med. sativa* and fixed nitrogen with the same efficiency as the wild type [72].

Table 1: Siderophores produced by rhizobia

Siderophore	Type	Rhizobial Species	References
-	hydroxamate	*Bradyrhizobium* spp.	[73]
vicibactin	hydroxamate	*Rhizobium leguminosarum*	[68]
-	hydroxamate	*Rhi. leguminosarum*	[74]
rhizobactin 1021	hydroxamate	*Ensifer meliloti* (formerly *Rhizobium meliloti*)	[75]
-	hydroxamate	*E. meliloti*	[44]
-	hydroxamate	*Rhizobium* spp.	[76]
ferric citrate	carboxylate	*Bradyrhizobium japonicum*	[77]
-	catecholate	*Bradyrhizobium* sp.	[78]
rhizobactin	carboxylate	*E. meliloti*	[79]
anthranilate	carboxylate	*E. meliloti* and *Rhi. leguminosarum*	[71, 80]
-	catecholate	*Rhi. leguminosarum*	[81]
-	catecholate	*Rhizobium* spp.	[82, 83]
-	catecholate	cowpea rhizobia	[84, 85]

3.2.3. Production of Ammonia by Rhizobia

Besides nitrogen fixation, some rhizobial strains can provide ammonia to environments *via* ammonification and urease activity. Ammonification can occur from the degradation of nitrogen compounds such as nitrate [86], casamino acid and trypticase peptone [87]. Urease is a nickel-containing, multi-subunit enzyme that catalyzes the hydrolysis of urea to ammonia and carbonic acid [88]. The 23 rhizobial isolates that nodulate either *Pha. vulgaris* or *V. radiata* produced ammonia from ammonification of peptone ranging between 70.00 ± 15.56 to 266.35 ± 23.83 µM. These isolates were observed to produce extracellular urease that ranged between 5.46 ± 1.02 to 28.33 ± 5.62 unit/ml supernatant [89].

3.2.4. Phosphate-Solubilizing Activity of Rhizobia

Phosphorus (P) is a major essential macronutrient for biological growth and development [90]. Inorganic phosphate as well as its organic esters and anhydrides are the predominant forms of phosphorus found in biological systems. These compounds are required metabolites in all living organisms [91]. Phosphorus may exist in 2 forms including soluble and insoluble forms. Phosphorus in insoluble forms, such as tricalcium phosphate $[Ca_3(PO_4)_2]$, iron phosphate $(FePO_4)$ and aluminium phosphate $(AlPO_4)$, are unavailable to living organisms including microorganisms and plants [92].

The concentration of soluble phosphorus in soil is usually very low, normally at levels of 1 ppm or less [93]. Phosphorus is usually applied to soil in the form of phosphatic fertilizers [94]. However, a large portion of soluble inorganic phosphate is rapidly immobilized soon after the application and becomes unavailable to plants [95]. The solubility of phosphorus is inhibited by the presence of iron or aluminium in acidic soils as well as calcium in neutral and alkaline soils [96]. Some microorganisms have the ability to convert insoluble forms of phosphorus to an accessible forms for plants. These specific groups of microorganisms are termed as "phosphate-solubilizing microorganisms" that play an important role in the solubilization of inorganic phosphates, then increasing in phosphorus uptake by the plants [97]. Several bacteria and fungi were evaluated for their mineral phosphate-solubilizing activity with various phosphorus sources such as tricalcium phosphate [90, 98, 99], iron phosphate [100] and aluminium phosphate [98]. Bacteria in different genera and species have been reported as "phosphate-solubilizing bacteria (PSB)". Several reports have examined the ability of different bacterial species to solubilize insoluble inorganic phosphate compounds, such as tricalcium phosphate, dicalcium phosphate, hydroxyapatite and rock phosphate [94]. Visual detection and even semiquantitative estimation of the phosphate-solubilizing ability of microorganisms have been possible using the plate screening method, which shows clearing zones around the microbial colonies in media containing insoluble mineral phosphates (mostly tricalcium phosphate or hydroxyapatite) as the sole phosphorus sources [101]. Phosphate-solubilizing activity of PSB was strongly associated with either the secretion of low molecular weight organic acids [97, 102] or the enzyme production [92, 96, 99, 102-104]. Organic acids can greatly increase the phosphate solubilization through their hydroxyl and carboxyl groups chelate the cations bond to phosphate, thereby coverting it into soluble forms [105]. Phosphatase is one of enzymes that release phosphorus from organic compounds by dephosphorylation of phospho-ester or phospho-anhydride bonds in organic matter [106]. Acid-, neutral- and alkaline phosphatases catalyze such a reaction at acid, neutral and alkaline conditions, respectively. Phosphatases play a major role in mineralization of organic forms of phosphorus in soil [96, 99, 103, 104]. The enhancement of acid phosphatase and alkaline phosphatase activity has been attributed to the phosphate solubilization for the phosphorus requirement of bacteria [102]. Recently, the application of PSB to increase phosphorus availability to plants has also been reported [102, 107, 108]. Phosphate-solubilizing ability is considered as one of the most important traits associated with plant phosphorus nutrition.

Rhizobia are ones of the most powerful phosphate solubilizers. The capacity of rhizobia to solubilize phosphate has received an attention in recent years. Phosphate-solubilizing ability has been reported for *Bradyrhizobium* spp., *Cupriavidus* spp., *Sinorhizobium* spp., *Mesorhizobium ciceri*, *Mesorhizobium mediterraneum*, *Mesorhizobium* spp., *Rhi. leguminosarum* and *Rhizobium* spp. [47, 92, 101, 102, 109-113]. The 56 rhizobial strains produced extracellular phosphatases in medium containing tricalcium phosphate as a sole phosphorus source. These strains produced extracellular acid phosphatase ranged between 2.81 ± 0.17 to 7.86 ± 0.26 milliunit/ml, extracellular neutral phosphatase ranged between 2.46 ± 0.14 to 5.88 ± 0.01 milliunit/ml and extracellular alkaline phosphatase ranged between 2.96 ± 0.28 to 12.65 ± 0.13 milliunit/ml [92]. There are reports on plant-growth promotion by rhizobia that have phosphate-solubilizing activity after their inoculation into soil or plant seeds. *Rhi. leguminosarum* with phosphate-solubilizing activity increased dry matter of lettuce and maize (corn) under field condition [48]. *Rhi. leguminosarum* colonized roots better than did the altered mutants defective in phosphate solubilization, suggesting that the phosphate solubilization might have an important role in rhizosphere competitiveness [49]. The strain of *Burkholderia cepacia*, showing no IAA production but displaying significant mineral phosphate solubilization and moderate phosphatase activity, improved the yields of banana, citrus, coffee, onion, potato and tomato [101]. Inoculation of 2 strains of *Rhi. leguminosarum* with phosphate-solubilizing ability improved the root colonization and the growth promotion as well as significantly increased the phosphorus concentration in lettuce and maize [47, 48]. *Mesorhizobium* and *Sinorhizobium* strains with phosphate-solubilizing activity could stimulate simultaneous increase in plant dry weight and phosphorus content of corn [102].

3.2.5. Production of 1-Aminocyclopropane-1-Carboxylate (ACC) Deaminase by Rhizobia

The enzyme 1-aminocyclopropane-1-carboxylate (ACC) deaminase hydrolyzes ACC (the immediate precursor of ethylene) to ammonia and α-ketobutyrate [114]. As ethylene is a phytohormone that inhibits nodulation in various legumes, the ACC-producing rhizobia are supposed to decrease ethylene level in plants. ACC deaminase might be the strategy utilized by *Rhizobium* to promote nodulation by adjusting ethylene levels in legumes [115].

A positive correlation between the *in vitro* ACC deaminase activity of bacteria and their stimulating effect on root elongation suggests that the utilization of ACC is an important bacterial trait determining root growth promotion [114]. ACC deaminase in *Rhi. leguminosarum* biovar viciae 128C53K enhanced the nodulation of *Pisum sativum* (pea), likely by modulating ethylene levels in the plant roots during the early stages of nodule development. The ACC deaminase structural gene, *acdS*, and its upstream regulatory gene from *Rhi. leguminosarum* biovar viciae 128C53K were introduced into *E. meliloti*, which does not produce this enzyme. The resulting ACC deaminase-producing *E. meliloti* strain showed 35% to 40% greater efficiency in nodulating *Med. sativa*. Furthermore, the ACC deaminase-producing *E. meliloti* strain was more competitive in nodulation than the wild-type strain, suggesting that the increased competitiveness might be related to the utilization of ACC as a nutrient within the infection threads [116]. Like other symbiotic genes, *acdS* gene of *Mes. loti* MAFF303099 is positively regulated by the NifA2 protein, a transcriptional activator of the nitrogenase system. The mode of gene expression suggests that *acdS* participates in the establishment and/or the maintenance of mature nodules by interfering with the production of ethylene [117].

REFERENCES

[1]   Pueppke SG, Broughton WJ. *Rhizobium* sp. strain NGR234 and *R. fredii* USDA 257 share exceptionally broad, nested host ranges. Mol Plant-Microbe Interact 1999; 12: 293-18.

[2]   Appunu C, Sen D, Singh MK, Dhar B. Variation in symbiotic performance of *Bradyrhizobium japonicum* strains and soybean cultivars under field conditions. J Cent Eur Agric 2008; 9: 185-90.

[3]   Appunu C, Reddy LML, Reddy CVCM, *et al.* Symbiotic diversity among acid-tolerant bradyrhizobial isolates with cowpea. J Agric Sci 2009; 4: 126-31.

[4]   Fening JO, Danso SKA. Variation in symbiotic effectiveness of cowpea bradyrhizobia indigenous to Ghanaian soils. Appl Soil Ecol 2002; 21: 23-29.

[5]   Jones DG. Symbiotic variation of *Rhizobium trifolii* with S.100 nomark white clover (*Trifolium repens* L.). J Sci Food Agric 2006; 14: 740-43.

[6]   Burdon JJ, Gibson AH, Woods MJ, Brockwell J. Variation in the effectiveness of symbiotic associations between native rhizobia and temperate Australian *Acacia*: within-species interactions. J Appl Ecol 1999; 36: 398-08.

[7]   Gross DC, Vidaver AK, Klucas RV. Plasmids, biological properties and efficacy of nitrogen fixation in *Rhizobium japonicum* strains indigenous to alkaline soils. J Gen Microbiol 1979; 114: 257-66.

[8]   Miller RW, Sirois JC. Relative efficacy of different alfalfa cultivar-*Rhizobium meliloti* strain combinations for symbiotic nitrogen fixation. Appl Envir Microbiol 1982; 43: 764-68.

[9]   Balatti PA, Pueppke SG. Nodulation of soybean by a transposon-mutant of *Rhizobium fredii* USDA 257 is subject to competitive nodulation blocking by other rhizobia. Plant Physiol 1990; 94: 1276-81.

[10]   Broughton WJ, Heycke N, Meyer ZA, Pankhurst CE. Plasmid-linked *nif* and *nod* genes in fast-growing rhizobia that nodulate *Glycine max*, *Psophocarpus tetragonolobus*, and *Vigna unguiculata*. Proc Natl Acad Sci USA. 1984; 81: 3093-97.

[11]   Somasegaran P, Martin RB. Symbiotic characteristics and *Rhizobium* requirements of a *Leucaena leucocephala* x *Leucaena diversifolia* hybrid and its parental genotypes. Appl Envir Microbiol 1986; 52: 1422-24.

[12]   Israel DW, Mathis JN, Barbour WM, Elkan GH. Symbiotic effectiveness and host-strain interactions of *Rhizobium fredii* USDA 191 on different soybean cultivars. Appl Envir Microbiol 1986; 51: 898-03.

[13]   Friedericks JB, Hagedorn C, Vanscoyoc SW. Isolation of *Rhizobium leguminosarum* (biovar *trifolii*) strains from Ethiopian soils and symbiotic effectiveness on African annual clover species. Appl Envir Microbiol 1990; 56: 1087-92.

[14]   Silva C, Vinuesa P, Eguiarte LE, *et al. Rhizobium etli* and *Rhizobium gallicum* nodulate common bean (*Phaseolus vulgaris*) in a traditionally managed Milpa Plot in Mexico: population genetics and biogeographic implications. Appl Envir Microbiol 2003; 69: 884-93.

[15]   Fuhrmann J. Symbiotic effectiveness of indigenous soybean bradyrhizobia as related to serological, morphological, rhizobitoxine, and hydrogenase phenotypes. Appl Envir Microbiol 1990; 56: 224-29.

[16]   Kleczkowska J. Mutations in symbiotic effectiveness in *Rhizobium trifolii* caused by transforming DNA and other agents. Microbiol 1965; 40: 377-83.

[17]   Toro N, Olivares J. Characterization of a large plasmid of *Rhizobium meliloti* involved in enhancing nodulation. Mol Gen Genet 1986; 202: 331-35.

[18]   Velazquez E, Mateos PE, Pedrero P, *et al.* Attenuation of symbiotic effectiveness by *Rhizobium meliloti* SAF22 related to the presence of a cryptic plasmid. Appl Envir Microbiol 1995; 61: 2033-36.

[19]   Lievens S, Goormachtig S, Den Herder J, *et al.* Gibberellins are involved in nodulation of *Sesbania rostrata.* Plant Physiol 2005; 139: 1366-79.

[20]   Dobert RC, Rood SB, Blevins DG. Gibberellins and the legume-*Rhizobium* symbiosis: I. endogenous gibberellins of lima bean (*Phaseolus lunatus* L.) stems and nodules. Plant Physiol 1992; 98: 221-24.

[21]   Davies PJ. The Plant Hormone Concept: Concentration, Sensitivity, and Transport. In: Davies PJ, Ed. Plant Hormones: Physiology, Biochemistry and Molecular Biology. Dordrecht, Kluwer Academic Publishers, 1995; pp. 13-18.

[22]   Lambrecht M, Okon Y, Broek AV, Vanderleyden J. Indole-3-acetic acid: a reciprocal signaling molecule in bacteria-plant interactions. Trends Microbiol 2000; 8: 298-00.

[23]   Patten CL, Glick BR. Bacterial biosynthesis of indole-3-acetic acid. Can J Microbiol 1996; 42: 207-20.

[24]   Patten CL, Glick BR. Role of *Pseudomonas putida* indole acetic acid in development of the host root system. Appl Envir Microbiol 2002; 68: 3795-01.

[25]   Hunter WJ. Indole-3-acetic acid production by bacteroids from soybean root nodules. Physiologia Plantarum 1989; 76: 31-36.

[26]   Pongsilp N, Nuntagij A. Genetic diversity and metabolites production of root-nodule bacteria isolated from medicinal legumes *Indigofera tinctoria*, *Pueraria mirifica* and *Derris elliptica* Benth. grown in different geographic origins across Thailand. Amer-Eur J Agric Envir Sci 2009; 6: 26-34.

[27]   Broughton WJ, Dilworth MJ. Control of leghaemoglobin synthesis in snake beans. Biochem J 1971; 125: 1075-80.

[28]   Carelli M, Gnocchi S, Fancelli S, *et al.* Genetic diversity and dynamics of *Sinorhizobium meliloti* populations nodulating different alfalfa cultivars in Italian soils. Appl Envir Microbiol 2000; 66: 4785-89.

[29]   Jameson PE. Cytokinins and auxins in plant-pathogen interactions: an overview. Plant Growth Regul 2000; 32: 369-80.

[30]   Basu PS, Ghosh AC. Indole acetic acid and its metabolism in root nodules of a monocotyledonous tree *Roystonea regia.* Curr Microbiol 1998; 37: 137-40.

[31]   Fukuhara H, Minakawa Y, Akao S, Minamisawa K. The involvement of indole-3-acetic acid produced by *Bradyrhizobium elkanii* in nodule formation. Plant Cell Physiol 1994; 35: 1261-65.

[32]   Mathesius U, Schlaman HRM, Spaink HP, *et al.* Auxin transport inhibition precedes root nodule formation in white clover roots and is regulated by flavonoids and derivatives of chitin oligosaccharides. Plant J 1998; 14: 23-34.

[33] Kittell BL, Helinski DR, Ditta GS. Aromatic aminotransferase activity and indole acetic acid production in *Rhizobium meliloti*. J Bacteriol 1989; 171: 5458-66.

[34] Kawaguchi M, Syono K. Excessive production of indole-3-acetic acid and its significance in studies of the biosynthesis of this regulator of plant growth and development. Plant Cell Physiol 1996; 37: 1043-48.

[35] Hull AK, Vij R, Celenza JL. *Arabidopsis* cytochrome P450s that catalyze the first step of tryptophan-dependent indole-3-acetic acid biosynthesis. Proc Natl Acad Sci USA. 2000; 97: 2379-84.

[36] Kobayashi M, Suzuki T, Fujita T, *et al.* Occurrence of enzymes involved in biosynthesis of indole-3-acetic acid from indole-3-acetonitrile in plant-associated bacteria, *Agrobacterium* and *Rhizobium*. Proc Natl Acad Sci USA. 1995; 92: 714-18.

[37] Minamisawa K, Ogawa K, Fukuhara H, Koga J. Indole pyruvate pathway for indole-3-acetic acid biosynthesis in *Bradyrhizobium elkanii*. Plant Cell Physiol 1996; 37: 449-53.

[38] Sekine M, Ichikawa T, Kuga N, *et al.* Detection of the IAA biosynthetic pathway from tryptophan *via* indole-3-acetamide in *Bradyrhizobium* spp. Plant Cell Physiol 1988; 29: 867-74.

[39] Kawaguchi M, Sekine M, Syono K. Isolation of *Rhizobium leguminosarum* biovar *viciae* variants with indole-3-acetamide hydrolase activity. Plant Cell Physiol 1990; 31: 449-55.

[40] Leelahawonge C, Nuntagij A, Pongsilp N. Factors influencing indole-3-acetic acid biosynthesis of root-nodule bacteria isolated from various leguminous plants. Thammasat Int J Sci Tech 2009; 14: 1-12.

[41] Bhattacharyya RN, Pati BR. Growth behaviour and indole acetic acid (IAA) production by a *Rhizobium* isolated from root nodules of *Alysicarpus vaginalis* DC. Acta Microbiol Immunol Hung 2000; 47: 41-51.

[42] Datta C, Basu PS. Indole acetic acid production by a *Rhizobium* species from root nodules of a leguminous shrub, *Cajanus cajan*. Microbiol Res 2000; 155: 123-27.

[43] Bhattacharyya RN, Basu PS. Bioproduction of indole acetic acid by a *Rhizobium* sp. from the root nodules of *Desmdium gangeticum* DC. Acta Microbiol Immunol Hung 1997; 44: 109-18.

[44] Arora NK, Kang SC, Maheshwari DK. Isolation of siderophore-producing strains of *Rhizobium meliloti* and their biocontrol potential against *Macrophomina phaseolina* that causes charcoal rot of groundnut. Curr Sci 2001; 81: 673-77.

[45] Ghosh S, Basu PS. Production and metabolism of indole acetic acid in roots and root nodules of *Phaseolus mungo*. Microbiol Res 2006; 161: 362-66.

[46] Williams MNV, Signer ER. Metabolism of tryptophan and tryptophan analogs by *Rhizobium meliloti*. Plant Physiol. 1990; 92: 1009-13.

[47] Chabot R, Antoun H, Kloepper JW, Beauchamp CJ. Root colonization of maize and lettuce by bioluminescent *Rhizobium leguminosarum* biovar phaseoli. Appl Envir Microbiol 1996; 62: 2767-72.

[48] Chabot R, Antoun H, Cescas MP. Growth promotion of maize and lettuce by phosphate-solubilizing *Rhizobium leguminosarum* biovar. *phaseoli*. Plant Soil 1996; 184: 311-21.

[49] Chabot R, Beauchamp CJ, Kloepper JW, Antoun H. Effect of phosphorus on root colonization and growth promotion of maize by bioluminescent mutants of phosphate-solubilizing *Rhizobium leguminosarum* biovar *phaseoli*. Soil Biol Biochem 1998; 30: 1615-18.

[50] Hoflich G, Wiehe W, Hecht-Buchholz C. Rhizosphere colonization of different crops with growth promoting *Pseudomonas* and *Rhizobium* bacteria. Microbiol Res 1995; 150: 139-47.

[51]   Wiehe W, Hoflich G. Survival of plant growth promoting rhizosphere bacteria in the rhizosphere of different crops and migration to noninoculated plants under field conditions in north-east Germany. Microbiol Res 1995; 150: 201-06.

[52]   Yanni YG, Rizk RY, Corich V, *et al.* Endorhizosphere colonization and growth promotion of indica and japonica rice varieties by *Rhizobium leguminosarum* bv. *trifolii.* Proceedings of the 15th Symbiotic Nitrogen Fixation Conference, North Carolina State University, Raleigh; USA. 1995.

[53]   Nimnoi P, Pongsilp N. Genetic diversity and plant-growth promoting ability of the indole-3-acetic acid (IAA) synthetic bacteria isolated from agricultural soil as well as rhizosphere, rhizoplane and root tissue of *Ficus religiosa* L., *Leucaena leucocephala* and *Piper sarmentosum* Roxb. Res J Agric Biol Sci 2009; 5: 29-41.

[54]   Kaneshiro T, Kwolek WF. Stimulated nodulation of soybeans by *Rhizobium japonicum* mutant (B-14075) that catabolizes the conversion of tryptophan to indol-3yl-acetic acid. Plant Sci 1985; 42: 141-46.

[55]   Bianco C, Defez R. Improvement of phosphate solubilization and *Medicago* plant yield by an indole-3-acetic acid-overproducing strain of *Sinorhizobium meliloti.* Appl Envir Microbiol 2010; 76: 4626-32.

[56]   Badenoch-Jones J, Summons RE, Djordjevic MA, *et al.* Mass spectrometric quantification of indole-3-acetic acid in *Rhizobium* culture supernatants: relation to root hair curling and nodule initiation. Appl Envir Microbiol 1982; 44: 275-80.

[57]   Hunter WJ. Influence of 5-methyltryptophan-resistant *Bradyrhizobium japonicum* on soybean root nodule indole-3-acetic acid content. Appl Envir Microbiol 1987; 53: 1051-55.

[58]   Bianco C, Defez R. *Medicago truncatula* improves salt tolerance when nodulated by an indole-3-acetic acid over-producing *Sinorhizobium meliloti* strain. J Exp Bot 2009; 60: 3097-07.

[59]   King RW, Evans LT. Gibberellins and flowering of grasses and cereals: prising open the lid of the "Florigen" black box. Annu Rev Plant Biol 2003; 54: 307-28.

[60]   Miransari M, Smith D. Rhizobial lipo-chitooligosaccharides and gibberellins enhance barley (*Hordeum vulgare* L.) seed germination. Biotechnol 2009; 8: 270-75.

[61]   Hedden P, Phillips AL. Gibberellin metabolism: new insights revealed by the genes. Trends Plant Sci 2000; 5: 523-30.

[62]   Yamaguchi S, Kamiya Y. Gibberellin biosynthesis: its regulation by endogenous and environmental signals. Plant Cell Physiol 2000; 41: 251-57.

[63]   Atzorn R, Crozier A, Wheeler CT, Sandberg G. Production of gibberellins and indole-3-acetic acid by *Rhizobium phaseoli* in relation to nodulation of *Phaseolus vulgaris* roots. Planta 1988; 175: 532-38.

[64]   Williams PM, de Mallorca MS. Abscisic acid and gibberellin-like substances in roots and root nodules of *Glycine max*. Plant Soil 1982; 65: 19-26.

[65]   Hotta K, Kim CY, Fox DT, Koppisch AT. Siderophore-mediated iron acquisition in *Bacillus anthracis* and related strains. Microbiol 2010; 156: 1918-25.

[66]   Stevens JB, Carter RA, Hussain H, *et al.* The *fhu* genes of *Rhizobium leguminosarum,* specifying siderophore uptake proteins: *fhuDCB* are adjacent to a pseudogene version of *fhuA.* Microbiol 1999; 145: 593-01.

[67]   Xiao R, Kisaalita WS. Iron acquisition from transferrin and lactoferrin by *Pseudomonas aeruginosa* pyoverdin. Microbiol 1997; 143: 2509-15.

[68] Dilworth MJ, Carson KC, Giles RGF, *et al. Rhizobium leguminosarum* bv. *viciae* produces a novel cyclic trihydroxamate siderophore, vicibactin. Microbiol 1998; 144: 781-91.

[69] Lynch D, O'Brien J, Welch T, *et al.* Genetic organization of the region encoding regulation, biosynthesis, and transport of rhizobactin 1021, a siderophore produced by *Sinorhizobium meliloti.* J Bacteriol 2001; 183: 2576-85.

[70] Reigh G, O'Connell M. Siderophore-mediated iron transport correlates with the presence of specific iron-regulated proteins in the outer membrane of *Rhizobium meliloti.* J Bacteriol 1993; 175: 94-02.

[71] Barsomian GD, Urzainqui A, Lohman K, Walker GC. *Rhizobium meliloti* mutants unable to synthesize anthranilate display a novel symbiotic phenotype. J Bacteriol 1992; 174: 4416-26.

[72] Chao TC, Becker A, Buhrmester J, *et al.* The *Sinorhizobium meliloti fur* gene regulates, with dependence on Mn(II), transcription of the *sitABCD* operon, encoding a metal-type transporter. J Bacteriol 2004; 186: 3609-20.

[73] Lesueur D, Diem HG, Meyer JM. Iron requirement and siderophore production in *Bradyrhizobium* strains isolated from *Acacia mangium.* J Appl Bacteriol 1993; 74: 675-82.

[74] Carson KC, Holliday S, Glenn AR, Dilworth MJ. Siderophore and organic acid production in root nodule bacteria. Arch Microbiol 1992; 157: 264-71.

[75] Persmark M, Pittman P, Buyer JS, *et al.* Isolation and structure of rhizobactin 1021, a siderophore from alfalfa symbiont *Rhizobium meliloti* 1021. J Am Chem Soc 1993; 115: 3950-56.

[76] Sridevi M, Mallaiah KV. Production of hydroxamate-type of siderophores by *Rhizobium* strains from *Sesbania sesban* (L.) Merr. Int J Soil Sci 2008; 3: 28-34.

[77] Guerinot ML, Meidl EJ, Plessner O. Citrate as a siderophore in *Bradyrhizobium japonicum.* J Bacteriol 1990; 172: 3298-03.

[78] Nambiar PTC, Sivaramakrishnan S. Detection and assay of siderophores in cowpea rhizobia (*Bradyrhizobium*) using radioactive Fe (^{59}Fe). Appl Microbiol Lett 1987; 4: 37-40.

[79] Smith MJ, Shoolery JN, Schwyn B, *et al.* Rhizobactin, a structurally novel siderophore from *Rhizobium meliloti.* J Am Chem Soc 1985; 107: 1739-43.

[80] Rioux CR, Jordan DC, Rattray JBM. Iron requirement of *Rhizobium leguminosarum* and secretion of anthranilic acid during growth on an iron-deficient medium. Arch Biochem 1986; 248: 175-82.

[81] Patel HN, Chakraborty RN, Desai SB. Isolation and partial characterization of phenolate siderophore from *Rhizobium leguminosarum* IARI102. FEMS Microbiol Lett 1988; 56: 131-34.

[82] Roy N, Bhattacharyya P, Chakrabartty PK. Iron acquisition during growth in an iron deficient medium by *Rhizobium* sp. isolated from *Cicer arietinum.* Microbiol 1994; 140: 2811-20.

[83] Sridevi M, Kumar KG, Mallaiah KV. Production of catechol-type of siderophores by *Rhizobium* sp. isolated from stem nodules of *Sesbania procumbens* (Roxb.) W and A. Res J Microbiol 2008; 3: 282-87.

[84] Jadhav RS, A. Desai A. Role of siderophore in iron uptake in cowpea *Rhizobium* GN1 (peanut isolate): possible involvement of iron repressible outer membrane proteins. FEMS Microbiol Lett 1994; 115: 185-89.

[85] Modi M, Shah KS, Modi VV. Isolation and characterisation of catechol-type siderophore from cowpea *Rhizobium* RA-1. Arch Microbiol 1985; 141: 156-58.

[86]   Hoffmann T, Frankenberg N, Marino M, Jahn D. Ammonification in *Bacillus subtilis* utilizing dissimilatory nitrite reductase is dependent on *resDE*. J Bacteriol 1998; 180: 186-89.

[87]   McSweeney CS, Palmer B, Bunch R, Krause DO. Isolation and characterization of proteolytic ruminal bacteria from sheep and goats fed the tannin-containing shrub legume *Calliandra calothyrsus*. Appl Envir Microbiol 1999; 65: 3075-83.

[88]   Morou-Bermudez E, Burne RA. Genetic and physiologic characterization of urease of *Actinomyces naeslundii*. Infect Immun 1999; 67: 504-12.

[89]   Chuntanom S, Pongsilp N. Environmental parameters affecting urease production and ammonification in *Phaseolus vulgaris*-nodulating rhizobia and *Vigna radiata*-nodulating rhizobia. Int J Microbiol Res 2011; 2: 222-32.

[90]   Hu X, Chen J, Guo J. Two phosphate- and potassium-solublizing bacteria isolated from Tianmu Mountain, Zhejiang, China. World J Microbiol Biotechnol 2006; 22: 983-90.

[91]   Wilson MA, Metcalf WW. Genetic diversity and horizontal transfer of genes involved in oxidation of reduced phosphorus compounds by *Alcaligenes faecalis* WM2072. Appl Envir Microbiol 2005; 71: 290-96.

[92]   Leelahawonge C, Pongsilp N. Phosphatase activities of root-nodule bacteria and nutritional factors affecting production of phosphatases by representative bacteria from three different genera. KMITL Sci Tech J 2009; 9: 65-83.

[93]   Goldstein AH. Involvement of the quinoprotein glucose dehydrogenase in the solubilization of exogenous phosphates by gram-negative bacteria. In: Torriani-Gorini A, Yagil E, Silver S, Eds. Phosphate in Microorganisms: Cellular and Molecular Biology. Washington DC, ASM Press, 1994; pp. 197–03.

[94]   Goldstein AH. Bacterial solubilization of mineral phosphates: historical perspective and future prospects. Am J Altern Agric 1986; 1: 51-57.

[95]   Dey KB. Phosphate Solubilizing Organisms in Improving Fertility Status. In: Sen SP, Palit P, Eds. Biofertilizers: Potentialities and Problems. Plant Physiology Forum. Calcutta, Naya Prokash, 1988; pp. 237-48.

[96]   Ponmurugan P, Gopi C. *In vitro* production of growth regulators and phosphate activity by phosphate solubilizing bacteria. Afr J Biotechnol 2006; 5: 348-50.

[97]   Chen YP, Rekha PD, Arun AB, *et al.* Phosphate solubilizing bacteria from subtropical soil and their tricalcium phosphate solubilzing abilities. Appl Soil Ecol 2006; 34: 33-41.

[98]   Illmer P, Schinner F. Solubilization of inorganic calcium phosphates-solubilization mechanisms. Soil Biol Biochem 1995; 27: 257-63.

[99]   Relwani L, Krishna P, Reddy MS. Effect of carbon and nitrogen sources on phosphate solubilization by a wild-type strain and UV-induced mutants of *Aspergillus tubingensis*. Curr Microbiol 2008; 57: 401-06.

[100] Jones D, Smith BFL, WilsonMJ, Goodman BA. Phosphate solubilizing fungi in a Scottish upland soil. Mycol Res 1991; 95: 1090-93.

[101] Rodriguez H, Fraga R. Phosphate solubilizing bacteria and their role in plant growth promotion. Biotechnol Adv 1999; 17: 319-39.

[102] Pantujit S, Pongsilp N. Phosphatase activity and effects of phosphate-solubilizing bacteria on yield and uptake of phosphorus in corn. World Appl Sci J 2010; 8: 429-35.

[103] Kim KY, Jordan D, McDonald GA. *Enterobacter agglomerans*, phosphate solubilizing bacteria, and microbial activity in soil: effect of carbon sources. Soil Biol Biochem 1998; 30: 995-03.

[104] Raghothama KG. Phosphate acquisition. Annu Rev Plant Physiol Plant Mol Biol 1999; 50: 665-93.

[105] Kpomblekou-A K, Tabatabai MA. Effect of organic acids on release of phosphorus from phosphate rocks. Soil Sci 1994; 158: 442-53.

[106] Rodriguez H, Fraga R, Gonzalez T, Bashan Y. Genetics of phosphate solubilization and its potential applications for improving plant growth-promoting bacteria. Plant Soil 2006; 287: 15-21.

[107] Han HS, Lee KD. Phosphate and potassium solubilizing bacteria effect on mineral uptake, soil availability and growth of eggplant. Res J Agric Biol Sci 2005; 1: 176-80.

[108] Zaidi A, Khan MS, Amil MD. Interactive effect of rhizotrophic microorganisms on yield and nutrient uptake of chickpea (*Cicer arietinum* L.). Europ J Agron 2003; 19: 15-21.

[109] Abril A, Zurdo-Pineiro JL, Peix A, *et al.* Solubilization of phosphate by a strain of *Rhizobium leguminosarum* bv. trifolii isolated from *Phaseolus vulgaris* in El Chaco Arido soil (Argentina). In: Velazquez E, Rodriguez-Barrueco C, Eds. First International Meeting on Microbial Phosphate Solubilization. Dordrecht, Springer, 2007; pp. 135-138.

[110] Halder AK, Chakrabartty PK. Solubilization of inorganic phosphate by *Rhizobium*. Folia Microbiol 1993; 38: 325-30.

[111] Halder AK, Mishra AK, Bhattacharyya P, Chakrabartty PK. Solubilization of rock phosphate by *Rhizobium* and *Bradyrhizobium*. J Gen Appl Microbiol 1990; 36: 81-92.

[112] Jia X, Knight JD, Leggett ME. Comparison of media used to evaluate *Rhizobium leguminosarum* biovar *viciae* for phosphate-solubilizing ability. Can J Microbiol 2009; 55: 910-15.

[113] Rivas R, Peix A, Mateos PF, *et al.* Biodiversity of populations of phosphate solubilizing rhizobia that nodulates chickpea in different Spanish soils. In: Velazquez E, Rodriguez-Barrueco C, Eds. First International Meeting on Microbial Phosphate Solubilization. Dordrecht, Springer, 2007; pp. 23-33.

[114] Belimov AA, Hontzeas N, Safronova VI, *et al.* Cadmium-tolerant plant growth-promoting bacteria associated with the roots of Indian mustard (*Brassica juncea* L. Czern.). Soil Biol Biochem 2005; 37: 241-50.

[115] Ma W, Guinel FC, Glick BR. *Rhizobium leguminosarum* biovar viciae 1-aminocyclopropane-1-carboxylate deaminase promotes nodulation of pea plants. Appl Envir Microbiol 2003; 69: 4396-02.

[116] Ma W, Charles TC, Glick BR. Expression of an exogenous 1-aminocyclopropane-1-carboxylate deaminase gene in *Sinorhizobium meliloti* increases its ability to nodulate alfalfa. Appl Envir Microbiol 2004; 70: 5891-97.

[117] Nukui N, Minamisawa K, Ayabe SI, Aoki T. Expression of the 1-aminocyclopropane-1-carboxylic acid deaminase gene requires symbiotic nitrogen-fixing regulator gene *nifA2* in *Mesorhizobium loti* MAFF303099. Appl Envir Microbiol 2006; 72: 4964-69.

CHAPTER 4

Genotypic Diversity of Rhizobia Assessed by Plasmid Profile

Neelawan Pongsilp[*]

Department of Microbiology, Faculty of Science, Silpakorn University, Nakhon Pathom, Thailand

Abstract: Rhizobia in diverse genera may contain small plasmids and/or large plasmids, termed as "megaplasmids". Megaplasmids are categorized as the symbiotic plasmid (pSym) or the cryptic plasmid. In some rhizobial strains, the symbiotic plasmid and the cryptic plasmid may coexist together. Although the symbiotic plasmids malinly encode symbiotic genes, some symbiotic plasmids also carry genes involved in various other functions. The symbiotic plasmids have been found in some genera of rhizobia including *Ensifer*, *Mesorhizobium*, *Rhizobium* and *Sinorhizobium*. Although the cryptic plasmids mainly carry genes encoding basic metabolic functions, utilization of nutrients and synthesis of cellular components, symbiotic-related genes are found to be located on some cryptic plasmids. Megaplasmid profiles of rhizobia vary greatly, both in size and number of plasmids. It is one of the characteristics used to examine genotypic diversity of rhizobia but it is less valuable for taxonomic proposes. The relation between megaplasmid profiles and some other genotypic characteristics has been reported.

Keywords: Cryptic plasmid, Genotypic diversity, Megaplasmid, Plasmid profile, Plasmid stability, Plasmid transfer, Replicator gene (*rep*), Rhizobia, Symbiosis island, Symbiotic plasmid (pSym).

4.1. SYMBIOTIC PLASMID AND CRYPTIC PLASMID OF RHIZOBIA

Usually, the genetic information in some genera of rhizobia is partitioned between chromosome and large plasmids. The symbiotic genes are parts of a complex genomic structure which contain a large amount of reiterated DNA sequences. In some genera such as *Ensifer*, *Sinorhizobium* and *Rhizobium*, nitrogen fixation (*nif* and *fix*) and nodulation (*nod*) genes are present on large plasmids (megaplasmids), termed as "the symbiotic plasmids" (pSym) [1]. Isolates showing an identical profile based on number and size of plasmids are characterized in the same plasmid profile. Plasmid profile analysis has been deemed important because

*Address correspondence to Neelawan Pongsilp: Department of Microbiology, Faculty of Science, Silpakorn University, Nakhon Pathom, Thailand; Tel: +66-34-245337; Fax: +66-34-245336-37; E-mail: neelawan@su.ac.th

many important traits, including host specificity, nodulation [2], nitrogen fixation [3], competitiveness of nodule occupancy [4], bacteriocin production [5] and hydrogen recycling [6], are encoded on rhizobial plasmids. In some genera such as *Azorhizobium*, *Bradyrhizobium* and most *Mesorhizobium* species, the symbiotic information is presented on a part of chromosome termed as "symbiosis island" [7, 8]. A 611-kilobases (kb) DNA segment was identified as a highly probable candidate of a symbiosis island of *Mesorhizobium loti* MAFF303099, 30 genes for nitrogen fixation and 24 genes for nodulation were assigned in this region [9]. Besides symbiotic genes, some other genes are also located on pSym. The pSym of *Ensifer meliloti* (formerly *Rhizobium meliloti*) L5-30 contains a region responsible for the catabolism of a substance, called rhizopine, present in *Medicago sativa* (alfalfa) root nodules elicited by this strain [10]. Melanin production is linked to pSym of *Rhizobium leguminosarum* [11, 12]. The symbiotic plasmid, pRP2JI, of *R. leguminosarum* also contains *psi* region, that inhibits exopolysaccharide synthesis, and *psr* region, that represses the transcription of *psi* gene [13].

The cryptic plasmids, coexist together with pSym, carry genes encoding functions such as exopolysaccharide synthesis, thiamine biosynthesis [14], melanin production [15], dicarboxylate transport [16], utilization of various carbon sources [17-19], catechol catabolism [20], utilization of nitrate, lipopolysaccharide production [17, 21], nitrous oxide reductase production [22], cell mobility, acid tolerance [23], hydrogenase uptake system, repABC family plasmid replication module, secretion system, autoinducer production and copper resistance [24]. It has been found that some cryptic plasmids contain symbiotic-related genes involved in nodulation and nitrogen-fixation functions [25], nodulation competitiveness [26, 27], nodule formation efficiency [15] and successful establishment of the symbiosis [21]. The large number, stability and sizes of cryptic plasmids in rhizobia indicate that they are likely to serve some functions in the soil environment [28]. The plasmid complement of *R. leguminosarum* strains was found to be stable under laboratory and field conditions, implying that the plasmids contribute advantageously to cell maintenance and/or replication [29]. The number of cryptic plasmids present in each strain varies from 1 to 10 [30]. The cryptic megaplasmid of a broad host range *Ensifer fredii* (formerly *Rhizobium* sp.) NGR234 contains host specific nodulation (*hsn*) genes for the hosts *Psophcucarpus tetragonolobus, Vigna unguiculata* (cowpea) and related legumes [31]. *E. meliloti* 41 was able to catabolize calystegins, compounds that are abundant only in root exudate of *Calystegium sepium*, and used them as a sole source of carbon and nitrogen. The calystegin catabolism (*cac*) gene in this strain is located on a self-transmissible plasmid, pRme4la, which is not essential to nitrogen-fixing symbiosis with legumes [32]. The 2 cryptic plasmids, pMLa and pMLb, of *Mes. loti* MAFF303099 carry genes for the ABC-transporter system, phosphate assimilation, two-component system, DNA replication and conjugation as well as only one gene for nodulation [9].

4.2. VARIATION IN MEGAPLASMID PROFILES OF RHIZOBIA

The megaplasmid profiles of rhizobia vary greatly, both in size (approximately 45 to 1,700 kb) and number of plasmids (1 to 7), as shown in Table **1**. The procedures desbribed by Baginsky *et al.* [33]; Eckhardt [34] and Wheatcroft *et al.* [35] have been used to visualize plasmids present in the strains. Plasmid profiles have been used to discriminate among rhizobial strains and seem to be conserved in genetically related isolates [36]. No plasmid was observed in some rhizobial species such as *Bradyrhizobium yuanmingense* [37]. Isolates sharing identical plasmid profiles were always grouped by repetitive extragenic palindromic-polymerase chain reaction (REP-PCR) fingerprinting [38]. The 45 *Rhizobium* strains nodulating *Hedysarum coronarium* L. (sulla), isolated from plants growing in different sites in Menorca Island and southern Spain, were classified into 19 plasmid profiles. No correlation was found between plasmid profile and geographical origin of the strains. Each of 42 strains possessed a single pSym, and in 3 strains, the spontaneous loss of pSym resulted in the loss of the nodulation capacity. In addition to pSym, 18 different cryptic plasmids were identified. A characteristic cryptic plasmid of >1,500 kb was present in all strains [39]. The 18 isolates of *Rhizobium gallicum* harbored 2 to 4 megaplasmids, of sizes ranging from approximately 50 to 1,500 kb, representing 7 distinct profiles. The 108 isolates of *Rhizobium etli* harbored 2 to 5 megaplasmids, of sizes ranging from approximately 125 to 700 kb, representing 26 plasmid profiles. All *R. gallicum* tested isolates contained a 1,500-kb megaplasmid, which was absent in *R. etli* population. Whereas all of the *R. etli* megaplasmid profiles included a 700-kb plasmid, which was not found in *R. gallicum* population [40]. The 56 isolates of *R. leguminosarum* biovar viciae were characterized by plasmid profile, total DNA restriction pattern and restriction fragment length polymorphism (RFLP) of chromosomal regions and of pSym. The cryptic plasmids defining classes were strongly associated with specific chromosomal backgrounds. In contrast, variations in pSym were not related with variations in the remaining genome [41]. The 3 strains of *Rhizobium* isolated from roots of *Derris elliptica* and *Pueraria mirifica* (white Kwao Kruea) harbored the same plasmid profile with 2 small plasmids (0.9 and 1.5 kb in size). One *Sinorhizobium* strain and 4 *Rhizobium* strains nodulating *D. elliptica* or *Indigofera tinctoria* (true indigo) harbored the same megaplasmid profile [42].

4.3. GENETIC DIVERSITY OF PLASMID REPLICATOR GENE, *repC*, IN RHIZOBIAL PLASMIDS

Besides size and number of plasmids present in the strains, distribution and sequences of some genes have been used for examining the genotypic diversity of rhizobia. For example, it was found that the diversity of plasmid replicator gene, *repC*, sequences from *R. leguminosarum* strains and *Ensifer* (formerly *Sinorhizobium*) strains was so great that some sequences were different from each other. PCR primers could amplify the *repC* fragment of about 750 base pairs (bp) from 39 out of 41 strains of *R. leguminosarum* and also from several *Ensifer* strains. Restriction endonuclease digestion showed that the PCR product from individual strains, though uniform in size, was often heterogeneous in sequence [29]. The *repC* gene was also found on pSym of *Rhizobium tropici* [30]. The replicator sequences (*repA*, *repB* and *repC*), which locate on pSym of *R. etli* CFN42, were found in pSym of other *R. etli* and in a cryptic plasmid of *R. tropici* [30]. Among 4 major groups of *repC*, 2 or more different *repC* sequences were present in each strain of *R. leguminosarum* biovar viciae. In all cases, the *repC* sequences were shown to be plasmid-associated. When *repC* sequences belonging to different groups were present in an isolate, they were usually associated with different plasmids [43]. The diversity of *repC* profile was examined for *R. leguminosarum* biovar viciae populations in arable soils and grass soils. The *repC* profiles were more diverse in arable soils than in grass soils. The result was consistent with the result obtained from 16S-23S ribosomal DNA internal transcribed spacer polymerase chain reaction-restriction fragment length polymorphism (16S-23S ITS PCR-RFLP) [44].

4.4. RHIZOBIAL PLASMID TRANSFER

As some plasmids are transmissable, they might be lost and regained in populations, rapidly change in copy number and potentially mutate [45]. Introduction of plants in monoculture, inoculation of *Rhizobium* at high densities and lack of selection acting on host plants for symbiotic compatibility with indigenous *Rhizobium* may alter rates of successful plasmid transfer across bacterial chromosomal groups [46]. A 370-kb cryptic plasmid, pRlo2037a, of *Mes. loti* (formerly *Rhizobium loti*) could be transferred into *R. leguminosarum*, *E. meliloti* and *Agrobacterium tumefaciens* [47]. Two self-transmissible pSym plasmids, pJB5JI encoding specific nodulation with *Pisum sativum* (pea) and pBRlAN encoding specific nodulation with *Trifolium repens* (white clover), could be transferred at high rates to various derivatives of *R. leguminosarum* and *Rhizobium trifolii*, resulting in the expression of the symbiotic information encoded by pSym plasmids. When these plasmids were transferred to more distantly related members including *E. meliloti* and *A. tumefaciens*, variation in the apparent expression of the symbiotic functions occurred. Neither plasmid could induce any of the *E. meliloti* derivatives to nodulate pea or white clover. However, both pSym plasmids could induce a particular mutant of *E. meliloti* to nodulate *Med. sativa*. Variations in the expression of the information also occurred when these plasmids were transferred to a strain of *A. tumefaciens*. Plasmid pBRlAN enabled the strain to nodulate white clover, whereas the same strain with pJB5JI could not nodulate pea [25]. In contrast, a strict association existed between pSym and chromosomal genetic groups was observed, suggesting a lack of successful transfer of pSym between major *Rhizobium* chromosomal types [46]. Some factors affecting conjugal transfer of rhizobial plasmids have been reported. Transfer of plasmid pRmeGR4b, a cryptic plasmid of *E. meliloti* GR4, was dependent upon the presence of another cryptic plasmid, pRmeGR4a, on the same donor cell. Plasmid transfer was increased when glutamate was the only nitrogen source, whereas conjugation was virtually undetectable on ammonium. Furthermore, in media containing both glutamate and ammonium as nitrogen sources, transfer was reduced almost 100-fold compared with that in media containing glutamate alone [48].

Table 1: Megaplasmid profiles of rhizobia

Rhizobial Strain	Plamid Size (kb)	References
Bradyrhizobium elkanii		
DASA 01244	270	[24]
DASA 01265	265	[24]
DASA 01304	220	[24]
USDA 46	280	[24]
USDA 94	90	[24]
USDA 227	285	[24]
Bradyrhizobium japonicum (formerly *Rhizobium japonicum*)		
DASA 01196	145 and 215	[24]
RJ10B	≈75, ≈110, ≈140 and ≈180	[49]
RJ12S	≈110, ≈140 and ≈180	[49]
RJ17W	≈140 and ≈180	[49]
RJ19FY	≈75, ≈140 and ≈180	[49]
RJ23A	≈130 and ≈150	[49]
USDA 135	75, 115 and 170	[24]
USDA 146	160	[24]
USDA 177	255	[24]
USDA 225	230	[24]
WA5099-1-1	≈125 and ≈175	[49]
3I1b117	≈200	[49]
3I1b135	≈72, ≈110, ≈170 and ≈175	[49]
Bradyrhizobium sp.		
BTAi1	229	[24]
Ensifer fredii		
NGR234 (formerly *Rhizobium* sp.)	536 (pSym) and > 1,700 or > 2,000	[50-52]
Ensifer medicae		
M102	300, ≈1,300 (pSym) and ≈1,700 (pSym)	[53]
T2	160, 300, ≈1,300 (pSym) and ≈1,700 (pSym)	[53]
T10	170, ≈1,300 (pSym) and ≈1,700 (pSym)	[53]
Ensifer meliloti (formerly *Rhizobium meliloti* and *Sinorhizobium meliloti*)		

Table 1: cont….

ATCC 9930	180, ≈1,300 (pSym) and ≈1,700 (pSym)	[53]
SM11	144 and 200	[54]
T15	45, 190, 220, ≈1,300 (pSym) and ≈1,700 (pSym)	[53]
T1580	100, ≈1,300 (pSym) and ≈1,700 (pSym)	[53]
T1607	150, 170, 200, ≈1,300 (pSym) and ≈1,700 (pSym)	[53]
1021	1,354 (pSym) and 1,683 (pSym)	[55]
2011	290 (pSym)	[56]
***Ensifer* sp.**		
T173	175 (pSym), >400, >400, ≈1,300 and ≈1,700	[53]
Mesorhizobium amorphae		
ACCC 19660, ACCC 19666, ACCC 19670, ACCC 19672, ACCC 19673, ACCC 19674, ACCC 19675 and B107	150 and 930 (pSym)	[57]
ACCC 19662, ACCC 19663, ACCC 19664, B102, B103, B108, B269, B272, B275, B279 and B288	550 and 930 (pSym)	[57]
ACCC 19665, ACCC 19676, B101, B104, B105, B106, B109, B110, B270, B271, B273, B274, B278, B280, B281, B283, B284, B286, B287, B289, B291, B292, SH109 and SH190012	930 (pSym)	[57]
Mesorhizobium loti (formerly *Rhizobium loti*)		
MAFF303099	≈ 208 and ≈352	[9]
NZP2037	360	[47]
NZP2213	180	[47]
***Mesorhizobium* sp.**		
HL56	550 and 930 (pSym)	[57]
B276, B277, B282, B285, B290, N206 and SH15003	930 (pSym)	[57]
Rhizobium etli		
CFN42	184, 194, 250, 371 (pSym), 505 and 642	[58, 59]
Miml, Mim7, Mimll and Mim12	185, 300, 300, 510, 600 (pSym) and 900	[36]
Miml-2	390, 450, 600 (pSym) and >1,000	[36]
Miml-4, Mim8-5 and Mim10-2	390, 600 (pSym) and 1,000	[36]

Table 1: cont….

Miml-1, Miml-3, Mim2, Mim3-4, Mim3-6, Mim4-4, Mim4-5, Mim5, Mim6, Mim6-2, Mim6-3, Mim8 and Mim10	390, 450, 600 (pSym) and 1,000	[36]
Mim3-5, Mim3-7, Mim4-2, Mim4-3, Mim5-2, Mim5-3, Mim5-4, Mim5-5 and Mim8-2	390, 510, 600 (pSym) and 1,000	[36]
Mim7-4	200, 350, 530, 630 (pSym) and 800	[36]
Mim7-5	100, 390, 450, 600 (pSym) and 1,000	[36]
Rhizobium galegae		
HAMBI 59A2, HAMBI 490, HAMBI 540, HAMBI 1147, HAMBI 1155, HAMBI 1174, HAMBI 1185, HAMBI 1187, HAMBI 1189, HAMBI 1460 and HAMBI 1461	>1,000	[60]
HAMBI 1143, HAMBI 1144, HAMBI 1145, HAMBI 1183 and HAMBI 1185	340 and >1,000	[60]
HAMBI 1146, HAMBI 1186, HAMBI 1190, HAMBI 1191, HAMBI 1207, HAMBI 1208 and HAMBI 1428	180 and >1,000	[60]
Rhizobium gallicum		
-	525 (pSym) or 550 (pSym)	[40]
Rhizobium huautlense		
S02, S03, S04, S07, S08, S09, S11, S12, S22, S24, S25, S27, S28, S29, S30, S31, S34, S35, S42, S43, S77, S78, S93, S94, S101, S102, S103, S104, S105, S106, S110, S111, S113, S114, S115, S116, S119,	400 and 900	[60]
S123, S124, S131, S132, S151-S165, S167, S168, S169 and S170		
S41	900	[60]
S172	220, 400 and 900	[60]
Rhizobium leguminosarum		
T83k3	155, 200, 255, 300, 440 and 480	[61]
UPM791	≈190, 280, 380 and >900	[33]
W8-7	≈120, ≈260, ≈300 (pSym), ≈490 and ≈550	[17]
W11-9	≈270 (pSym), ≈700 and ≈1,000	[17]
W14-2	≈230, ≈260, ≈400 and ≈700 (pSym)	[17]
3841	147, 151, 352, 488 (pSym), 684 and 870	[58, 62]

Table 1: cont….

Rhizobium tropici		
CFN 299	185, 225, 410 and >1,000	[60]
USDA 2813	<190, 190, 280 (pSym) and >900	[33]
USDA 2822	<190 and 280 (pSym)	[33]
USDA 2838, USDA 2840 and USDA 9030	≈190 and 280 (pSym)	[33]
Rhizobium **spp.**		
ACCC 19667 and ACCC 19677	440, 490, 610 and 930	[57]
DASA 57010, DASA 57027, DASA 68020 and DASA 68025	≈85, ≈170, ≈300 and >920	[42]
OR191, T136 and T1155	140, <1,300 (pSym) and >1,700	[53]
T1470	120, 280, <1,300 (pSym) and >1,700	[53]
T1473	120, 140, 280, <1,300 (pSym) and >1,700	[53]
UPM-HcN1	≈145, ≈170, ≈210, ≈400 (pSym) and >1,500	[39]
UPM-Hc8	≈175, ≈200, ≈340 (pSym) and >1,500	[39]
UPM-Hc15	≈110, ≈210, ≈240, ≈375 (pSym) and >1,500	[39]
UPM-Hc18	≈210, ≈235, ≈400 (pSym) and >1,500	[39]
UPM-Hc22	≈130, ≈170, ≈300, ≈370 and >1,500	[39]
UPM-Hc23	≈45, ≈100, ≈110, ≈170, ≈325, ≈530 (pSym) and >1,500	[39]
UPM-Hc24	≈180, ≈285 (pSym) and > 1,500	[39]
UPM-Hc27	≈210, ≈400 (pSym) and >1,500	[39]
UPM-Hc29	≈160, ≈285 (pSym) and > 1,500	[39]
UPM-Hc30.21	≈130, ≈180, ≈340 (pSym), >460 and >1,500	[39]
UPM-Hc42	≈65, ≈180, ≈340 (pSym) and >1,500	[39]

Table 1: cont....

UPM-Hc44	≈45, ≈100, ≈160, ≈460 (pSym) and >1,500	[39]
UPM-Hc47	≈145, ≈180, ≈210, ≈255, ≈285 (pSym) and > 1,500	[39]
UPM-Hc48	≈170, ≈285 (pSym) and > 1,500	[39]
UPM-Hc50	≈285 (pSym) and > 1,500	[39]
UPM-Hc100	≈180, ≈375 (pSym) and >1,500	[39]
UPM-Hc121	≈400 (pSym) and >1,500	[39]
UPM-Hc122	≈180, ≈400 (pSym) and >1,500	[39]
UPM-Hc133	≈180, ≈210, ≈340 (pSym) and >1,500	[39]
Sinorhizobium **sp.**		
DASA 68012	≈85, ≈170, ≈300 and >920	[42]

pSym: the symbiotic plasmid which carries genes essential for nodulation and nitrogen fixation.
Sizes [in kilobases (kb)] were calculated from those reported in megadalton (MDal).

REFERENCES

[1]   Rogel MA, Hernandez-Lucas I, Kuykendall KD, *et al.* Nitrogen-fixing nodules with *Ensifer adhaerens* harboring *Rhizobium tropici* symbiotic plasmids. Appl Envir Microbiol 2001; 67: 3264-68.

[2]   Johnston AWB, Beynon JL, Buchanan-Wollaston AV, *et al.* High frequency transfer of nodulating ability between strains and species of *Rhizobium.* Nature (London) 1978; 276: 634- 36.

[3]   Nuti MP, Lepidi AA, Prakash RK, *et al.* Evidence for nitrogen fixation genes on indigenous *Rhizobium* plasmids. Nature (London) 1979; 282: 533-35.

[4]   Brewin NJ, Wood EA, Young JPW. Contribution of the symbiotic plasmid to the competitiveness of *Rhizobium leguminosarum.* J Gen Microbiol 1983; 129: 2973-77.

[5]   Hirsch PR. Plasmid-determined bacteriocin production by *Rhizobium leguminosarum.* J Gen Microbiol 1979; 113: 219-28.

[6]   Brewin NJ, de Jong TM, Phillips DA, Johnston AWB. Co-transfer of determinants of hydrogenase activity and nodulation ability in *Rhizobium leguminosarum.* Nature (London) 1980; 288: 77-79.

[7]   Flores M, Morales L, Avila M, *et al.* Diversification of DNA sequences in the symbiotic genome of *Rhizobium etli.* J. Bacteriol 2005; 187; 7185-92.

[8]   Sawada H, Kuykendall LD, Young JM. Changing concepts in the systematics of bacterial nitrogen-fixing legume symbionts. J Gen Appl Microbiol 2003; 49: 155–79.

[9]   Kaneko T, Nakamura Y, Sato S, *et al.* Complete genome structure of the nitrogen-fixing symbiotic bacterium *Mesorhizobium loti.* DNA Res 2000; 6: 331-38.

[10]   Murphy PJ, Heycke N, Banvalvi Z, *et al.* Genes for the catabolism and synthesis of an opine-like compound in *Rhizobium meliloti* are closely linked and on the pSym plasmid. Proc Natl Acad Sci USA. 1987; 84: 493-97.

[11] Borthakur D, Lamb JW, Johnston AWB. Identification of two classes of *Rhizobium phaseoli* genes required for melanin synthesis, one of which is required for nitrogen fixation and activates the transcription of the other. Mol Gen Genet 1987; 207: 155-60.

[12] Hawidns FKL, Johnston AWB. Transcription of a *Rhizobium leguminosarum* biovar *phaseoli* gene needed for melanin synthesis is activated by *nifA* of *Rhizobium* and *Klebsiella pneumoniae*. Mol Microbiol 1988; 2: 331-37.

[13] Mimmack ML, Hong GF, Johnston AWB. Sequence and regulation of *psrA,* a gene on the Sym plasmid of *Rhizobium leguminosarum* biovar *phaseoli* which inhibits transcription of the *psi* genes. Microbiol 1994; 140: 455-61.

[14] Finan TM, Kunkel B, de Vos GF, Signer ER. Second symbiotic megaplasmid in *Rhizobium meliloti* carrying exopolysaccharide and thiamine synthesis genes. J Bacteriol 1986; 167: 66-72.

[15] Sanjuan J, Olivares J. Implication of *nifA* in regulation of genes located on a *Rhizobium meliloti* cryptic plasmid that affect nodulation efficiency. J Bacteriol 1989; 171: 4154-61.

[16] Watson RJ, Chan YK, Wheatcroft R, *et al. Rhizobium meliloti* genes required for C4-dicarboxylate transport and nitrogen fixation are located on a megaplasmid. J Bacteriol 1988; 170: 927-34.

[17] Baldani JI, Weaver RW, Hynes MF, Eardly BD. Utilization of carbon substrates, electrophoretic enzyme patterns, and symbiotic performance of plasmid-cured clover rhizobia. Appl Envir Microbiol 1992; 58: 2308-14.

[18] Charles TC, Singh RS, Finan TM. Lactose utilization and enzymes encoded by megaplasmids in *Rhizobium meliloti* SU47: implications for population studies. J Gen Microbiol 1990; 136: 2497-02.

[19] Charles TC, Finan TM. Analysis of a 1600-kilobase *Rhizobium meliloti* megaplasmid using defined deletions generated *in vivo*. Genetics 1991; 127: 5-20.

[20] Gajendiran N, Mahadevan A. Plasmid-borne catechol dissimilation in *Rhizobium* sp. FEMS Microbiol Ecol 1990; 73: 125-130.

[21] Brom S, de los Santos AG, Stepkowsky T, *et al.* Different plasmids of *Rhizobium leguminosarum* bv. phaseoli are required for optimal symbiotic performance. J Bacteriol 1992; 174: 5183-89.

[22] Chan Y, Wheatcroft R. Detection of a nitrous oxide reductase structural gene in *Rhizobium meliloti* strains and its location on the *nod* megaplasmid of JJ1c10 and SU47. J Bacteriol 1993; 175: 19-26.

[23] Chen H, Gartner E, Rolfe BG. Involvement of genes on a megaplasmid in the acid-tolerant phenotype of *Rhizobium leguminosarum* biovar trifolii. Appl Envir Microbiol 1993; 59: 1058-64.

[24] Cytryn E, Jitacksorn S, Giraud E, Sadowsky MJ. Insights learned from pBTAi1, a 229-kb accessory plasmid from *Bradyrhizobium* sp. strain BTAi1 and prevalence of accessory plasmids in other *Bradyrhizobium* sp. strains. ISME J 2008; 2: 158-70.

[25] Djordjevic MA, Zurkowski W, Shine J, B.G. Rolfe BG. Sym plasmid transfer to various symbiotic mutants of *Rhizobium trifolii, R. leguminosarum*, and *R. meliloti*. J Bacteriol 1983; 156: 1035-45.

[26] Martinez-Romero E, Rosenblueth M. Increased bean (*Phaseolus vulgaris* L.) nodulation competitiveness of genetically modified *Rhizobium* strains. Appl Envir Microbiol 1990; 56: 2384-88.

[27]   Toro N, Olivares J. Analysis of *Rhizobium meliloti sym* mutants obtained by heat treatment. Appl Envir Microbiol 1986; 51: 1148-50.

[28]   Shaw PD. Plasmid Ecology. In: Kosuge T, Nester EW, Eds. Plant-Microbe Interactions: Molecular and Genetics Perspectives, Vol. 2. New York, Macmillan Publishing Co., 1987; pp 3-39.

[29]   Turner SL, Rigottier-Gois L. Power RS, *et al.* Diversity of *repC* plasmid-replication sequences in *Rhizobium leguminosarum*. Microbiol 1996; 172: 1705-13.

[30]   Ramirez-Romero MA, Bustos P, Girard L, *et al.* Sequence, localization and characteristics of the replicator region of the symbiotic plasmid of *Rhizobium etli*. Microbiol 1997; 143: 2825-31.

[31]   Broughton WJ, Wong CH, Lewin A, *et al.* Identification of *Rhizobium* plasmid sequences involved in recognition of *Psophocarpus, Vigna,* and other legumes. J Cell Biol 1986; 102: 1173-82.

[32]   Tepfer D, Goldmann A, Pamboukdjian N, *et al.* A plasmid of *Rhizobium meliloti* 41 encodes catabolism of two compounds from root exudate of *Calystegium sepium*. J Bacteriol 1988; 170: 1153-61.

[33]   Baginsky C, Brito B, Imperial J, *et al.* Diversity and evolution of hydrogenase systems in rhizobia. Appl Envir Microbiol 2002; 68: 4915-24.

[34]   Eckhardt T. A rapid method for the identification of plasmid deoxyribonucleic acid in bacteria. Plasmid 1978; 1: 584-88.

[35]   Wheatcroft R, McRae DG, Miller RW. Changes in the *Rhizobium meliloti* genome and the ability to detect supercoiled plasmids during bacteroid development. Mol Plant-Microbe Interact 1990; 3: 9-17.

[36]   Wang ET, Rogel MA, Santos AG, *et al. Rhizobium etli* bv. mimosae, a novel biovar isolated from *Mimosa affinis*. Int J Syst Bacteriol 1999; 49: 1479-91.

[37]   Yao ZY, Kan FL, Wang ET, *et al.* Characterization of rhizobia that nodulate legume species of the genus *Lespedeza* and description of *Bradyrhizobium yuanmingense* sp. nov. Int J Syst Evol Microbiol 2002; 52: 2219-30.

[38]   Laguerre G, Louvrier P, Allard M, Amarger N. Compatibility of rhizobial genotypes within natural populations of *Rhizobium leguminosarum* biovar viciae for nodulation of host legumes. Appl Envir Microbiol 2003; 69: 2276-83.

[39]   Mozo T, Cabrera E, Ruiz-Argueso T. Diversity of plasmid profiles and conservation of symbiotic nitrogen fixation genes in newly isolated *Rhizobium* strains nodulating sulla (*Hedysarum coronarium* L.). Appl Envir Microbiol 1988; 54: 1262-67.

[40]   Silva C, Vinuesa P, Eguiarte LE, *et al. Rhizobium etli* and *Rhizobium gallicum* nodulate common bean (*Phaseolus vulgaris*) in a traditionally managed Milpa Plot in Mexico: population genetics and biogeographic implications. Appl Envir Microbiol 2003; 69: 884-93.

[41]   Laguerre G, Mazurier SI, Amarger N. Plasmid profiles and restriction fragment length polymorphism of *Rhizobium leguminosarum* bv. *viciae* in field populations. FEMS MIcrobiol Lett 1992; 101: 17-26

[42]   Pongsilp N, Nuntagij A. Genetic diversity and metabolites production of root-nodule bacteria isolated from medicinal legumes *Indigofera tinctoria, Pueraria mirifica* and *Derris elliptica* Benth. grown in different geographic origins across Thailand. Amer-Eur J Agric Envir Sci 2009; 6: 26-34.

[43] Rigottier-Gois L, Turner SL, Young JPW, Amarger N. Distribution of *repC* plasmid-replication sequences among plasmids and isolates of *Rhizobium leguminosarum* bv. *viciae* from field populations. Microbiol 1998; 144: 771-80.

[44] Palmer KM, Young JPW. Higher diversity of *Rhizobium leguminosarum* biovar viciae populations in arable soils than in grass soils. Appl Envir Microbiol 2000; 66: 2445-50.

[45] Modi RI, Adams J. Coevolution in bacterial-plasmid populations. Evol 1991; 45: 656-67.

[46] Wernegreen JJ, Harding EE, Riley MA. *Rhizobium* gone native: unexpected plasmid stability of indigenous *Rhizobium leguminosarum*. Proc Natl Acad Sci USA. 1997; 94: 5483-88.

[47] Pankhurst CE, Broughton WJ, Wieneke U. Transfer of an indigenous plasmid of *Rhizobium loti* to other rhizobia and *Agrobacterium tumefaciens*. J Gen Microbiol 1983; 129: 2535-43.

[48] Herrera-Cervera J, Olivares J, Sanjuan J. Ammonia inhibition of plasmid pRmeGR4a conjugal transfer between *Rhizobium meliloti* strains. Appl Envir Microbiol 1996; 62: 1145-50.

[49] Gross DC, Vidaver AK, Klucas RV. Plasmids, biological properties and efficacy of nitrogen fixation in *Rhizobium japonicum* strains indigenous to alkaline soils. J Gen Microbiol 1979; 114: 257-66.

[50] Flores M, Mavingui P, Girard L, *et al.* 1998. Three replicons of *Rhizobium* sp. strain NGR234 harbor symbiotic gene sequences. J Bacteriol 1998; 180: 6052-53.

[51] Marvingui P, Flores M, Guo X, *et al.* Dynamics of genome architecture in *Rhizobium* sp. strain NGR234. J. Bacteriol 2002; 184: 171-76.

[52] Viprey V, Rosenthal A, Broughton WJ, Perret X. Genetic snapshots of the *Rhizobium* species NGR234 genome. Genome Biol 2000; 1: 0014.1-17

[53] Bromfield ESP, Tambong JT, Cloutier S, *et al. Ensifer*, *Phyllobacterium* and *Rhizobium* species occupy nodules of *Medicago sativa* (alfalfa) and *Melilotus alba* (sweet clover) grown at a Canadian site without a history of cultivation. Microbiol 2010; 156: 505-20.

[54] Stiens M, Schneiker S, Keller M, *et al.* Sequence analysis of the 144-kilobase accessory plasmid pSmeSM11a, isolated from a dominant *Sinorhizobium meliloti* strain identified during a long-term field release experiment. Appl Envir Microbiol 2006; 72: 3662-72.

[55] Galibert F, Finan TM, Long SR, *et al.* The composite genome of the legume symbiont *Sinorhizobium meliloti*. Science 2001; 293: 668-72.

[56] Renalier MH, Batut J, Ghai J, *et al.* A new symbiotic cluster on the pSym megaplasmid of *Rhizobium meliloti* 2011 carries a functional *fix* gene repeat and a *nod* locus. J Bacteriol 1987; 169: 2231-38.

[57] Wang ET, van Berkum P, Sui XH, *et al.* Diversity of rhizobia associated with *Amorpha fruticosa* isolated from Chinese soils and description of *Mesorhizobium amorphae* sp. nov. Int J Syst Bacteriol 1999; 49: 51-65.

[58] Crossman LC, Castillo-Ramirez S, McAnnula C, *et al.* A common genomic framework for a diverse assembly of plasmids in the symbiotic nitrogen fixing bacteria. PLoS One 2008; 3: e2567.

[59] Gonzalez V, Santamaria RI, Bustos P, *et al.* The partitioned *Rhizobium etli* genome: genetic and metabolic redundancy in seven interacting replicons. Proc Natl Acad Sci USA. 2006; 103: 3834-39.

[60] Wang ET, van Berkum P, Beyene D, *et al. Rhizobium huautlense* sp. nov., a symbiont of *Sesbania herbacea* that has a close phylogenetic relationship with *Rhizobium galegae*. Int J Syst Bacteriol 1998; 48: 687-99.

[61] Ruiz-Sainz JE, Beringer JE, Gutierrez-Navarro AM. Effect of the fungicide captafol on the survival and symbiotic properties of *Rhizobium trifolii*. J Appl Bacteriol 1984; 57: 361-67.

[62] Young JPW, Crossman LC, Johnston AWB, *et al*. The genome of *Rhizobium leguminosarum* has recognizable core and accessory components. Genome Biol 2006; 7: R34.

CHAPTER 5

Genotypic Diversity of Rhizobia Assessed by Polymerase Chain Reaction (PCR) Fingerprinting

Neelawan Pongsilp[*]

Department of Microbiology, Faculty of Science, Silpakorn University, Nakhon Pathom, Thailand

Abstract: The molecular techniques have been widely used for typing and assessing the genetic diversity of microorganisms. The techniques based on polymerase chain reaction (PCR) fingerprinting have been employed to examine genotypic diversity of rhizobial populations and to discriminate among rhizobial strains. These techniques include random amplified polymorphic DNA (RAPD), two-primers RAPD (TP-RAPD), repetitive sequence based PCR (rep-PCR) and amplified fragment length polymorphism (AFLP). The 3 main techniques of rep-PCR are enterobacterial repetitive intergenic consensus (ERIC)-PCR, repetitive extragenic palindromic (REP)-PCR and BOX-PCR. While RAPD uses a single primer to amplify the segments of DNA randomly throughout the genome, rep-PCR uses pairs of primers (for ERIC- and REP-PCR) or a single primer (for BOX-PCR) to amplify the intervals between conserved repeated sequences present in genome. In AFLP, total genomic DNA is digested and then ligated to oligonucleotide adapters. A pair of primer is used to amplify the product from restriction. RAPD, rep-PCR and AFLP are suitable for distinguishing strains at species or below levels but they are less valuable for taxonomic purpose. TP-RAPD has been developed for taxonomy purpose as the patterns of strains in the same species have been found to be identical. The TP-RAPD patterns supported the proposal of novel species of rhizobia.

Keywords: Amplified fragment length polymorphism (AFLP), BOX-polymerase chain reaction (BOX-PCR), Enterobacterial repetitive intergenic consensus-polymerase chain reaction (ERIC-PCR), Genotypic diversity, Polymerase chain reaction (PCR) fingerprinting, Random amplified polymorphic DNA (RAPD), Repetitive extragenic palindromic-polymerase chain reaction (REP-PCR), Repetitive sequence based polymerase chain reaction (rep-PCR), Rhizobia, Two-primers random amplified polymorphic DNA (TP-RAPD).

---

*Address correspondence to Neelawan Pongsilp: Department of Microbiology, Faculty of Science, Silpakorn University, Nakhon Pathom, Thailand; Tel: +66-34-245337; Fax: +66-34-245336-37; E-mail: neelawan@su.ac.th

5.1. RANDOM AMPLIFIED POLYMORPHIC DNA (RAPD) ANALYSIS OF RHIZOBIA

5.1.1. Principle of RAPD

The polymerase chain reaction (PCR) technique is a method for amplifying a target sequence of DNA. Specificity is based on the use of 2 nucleotide primers that hybridize to complementary sequences on opposite strand of DNA and flank the target sequence. The DNA sample is first heated to separate the 2 DNA strands, the primers are allowed to bind the DNA, and each strand is copied by a DNA polymerase, starting at a primer site. The 2 DNA strands each serve as a template for the synthesis of new DNA from the primers. Repeated cycles of heat denaturation, annealing of the primers to their complementary sequences and extension of the annealed primers with DNA polymerase result in the exponential amplification of DNA segments of defined length. Random amplified polymorphic DNA (RAPD) technique is a type of PCR, in which a single primer, termed as "arbitary primer", anneals complementary DNA and amplifies the segments of DNA randomly throughout the genome. A single arbitrary primer is a short oligonucleotide (8-12 base pairs (bp) long with 60% to 70% guanine-cytosine (GC) content) that acts as both forward primer and reverse primer in PCR. The approach relies upon genetic variation among genomes for the relative locations complementary to a primer, which results in PCR products varying in number and molecular size. As the length of a primer increases, the genomic target sites should decrease because there is a less chance of finding perfect or almost perfect homologies between the target sites and a longer primer [1]. Arbitary primers used in RAPD analysis of rhizobia are shown in Table **1**. RAPD reaction is always carried out at low annealing temperatures (generally 34°C to 38°C). The low annealing temperature results in a more chance of primer annealing. Except for RAPD primers targeting small subunit ribosomal RNA gene (16S rDNA) sequence (such as the primer 879F), the annealing temperatures are relatively high (typically 50°C or 55°C). The typical PCR conditions are as follows: first denaturing at 94°C for 1 min; followed by 45 cycles of denaturing at 94°C for 1 min, annealing at 36°C for 1 min and extension at 72°C for 2 min; with final extension at 72°C for 1 min [2]. This kind of primer yielded DNA patterns that allow to discriminate at species or subspecies levels [3]. In general, PCR products vary in length from approximately 100 bp to 3.0 kilobases (kb), and in number of bands from 1 to 15. Separation of PCR products by agarose gel electrophoresis reveals the polymorphism termed as "RAPD fingerprint" or "RAPD profile". The presence or absence of each band is scored and the resultant binary matrix is used to construct a dendrogram. The unweighted pair group method with arithmetic mean (UPGMA) dendrogram can be constructed using the software programs such as the Image Master 1D Elite Software version 5.20 (Amersham, UK) [2].

5.1.2. RAPD for Examining Genotypic Diversity of Rhizobia

The RAPD patterns are strain dependent and therefore, they are useful in analyzing the intra-specific diversity of rhizobial populations [4]. RAPD has been used to measure the genetic variability and to determine the genetic relationships of rhizobial populations. The usefulness of RAPD analysis in the characterization of *Bradyrhizobium* strains has been reported [2, 5-7]. Paffetti *et al.* [8] demonstrated the considerable level of genetic diversity among *Ensifer meliloti* (formerly *Rhizobium meliloti*) strains which were phenotypically indistinguishable. RAPD discriminated slightly better among *Ens. meliloti* strains than did enterobacterial repetitive intergenic consensus (ERIC)-PCR. This might be explained by minor changes in RAPD primer binding sites which are under no constraints to sequence conservation among closely related strains [9]. de Oliveira *et al.* [10] showed the great genetic heterogeneity among *Rhizobium leguminosarum* biovar phaseoli and *Rhizobium tropici* strains. The 19 rhizobial strains that nodulate *Cicer arietinum* (chickpea) were separated into 4 groups according to the RAPD fingerprinting, in which groups I and II might be subspecies of *Mesorhizobium mediterraneum*, group III was a subspecies of *Mesorhizobium tianshanense* and group IV was a subspecies of *Mesorhizobium ciceri* [3]. A total of 215 rhizobial isolates from 3 medicinal legumes, including *Derris elliptica*, *Indigofera tinctoria* (true indigo) and *Pueraria mirifica* (white Kwao Kruca) naturally growing in 16 provinces of Thailand, were evaluated for DNA polymorphism using RAPD analysis. The isolates generated 92 identical RAPD profiles, indicating highly significant genetic diversity among isolates from distinct geographic areas. It was also found that the genetic diversity was affected slightly by the host plants rather than the geographic origins [2]. RAPD fingerprint patterns divided 179 bean isolates into 6 groups in which 5 of the groups were placed within *Rhizobium lusitanum* and the other group was placed within *R. tropici* [11].

5.1.3. Advantages and Limitations of RAPD

Besides being simpler and cheaper, RAPD is as effective as the more labor intensive restriction fragment length polymorphism (RFLP) for establishing the genetic relationships and identifying *Rhizobium* strains [10, 12]. A dendrogram can be constructed from RAPD patterns to present the genetic relatedness among rhizobial populations. Furthermore, many arbitrary primers provide more alternatives which are suitable for any particular population. A significant limitation is the identity of each product across the different genome templates. Products of identical molecular size may not represent the identical region in each genome but are the same size by mere coincidence, therefore dendrograms constructed may not reflect accurately the genetic relationships among genomes [13]. RAPD does not permit the investigation of the diversity of symbiotic plasmids among chromosomally closely related strains [12]. Another limitation is that RAPD is sensitive to several factors such as selection of primer, PCR mixture and PCR cycling condition, therefore the results should not be comparable among different laboratories.

Table 1: Arbitary primers used in RAPD analysis of rhizobia

Primer and Sequence (5' to 3')	Rhizobial Strain	References
CC1 (AGC AGC GTG G)	*Rhizobium leguminosarum* strains	[14]
GLA-02 (TGC CGA GCT G) GLA-04 (AAT CGG GCT G) GLA-05 (AGG GGT CTT G) GLA-07 (GAA ACG GGT G) GLA-08 (GTG ACG TAG G) GLA-09 (GGG TAA CGC C) GLA-10 (GTG ATC GCA G) GLA-11 (CAA TCG CCG T) GLA-12 (TCG GCG ATA G) GLA-13 (CAG CAC CCA C) GLA-14 (TCT GTG CTG G) GLA-15 (TTC CGA ACC C) GLA-16 (AGC CAG CGA A) GLA-18 (AGG TGA CCG T)	rhizobial strains	[15]

Table 1: cont...

GLA-19 (CAA ACG TCG G) GLA-20 (GTT GCG ATC C) GLC-01 (TTC GAG CCA G) GLC-02 (GTG AGG CGT C) GLC-05 (GAT GAC CGC C) GLC-06 (GAA CGG ACT C) GLC-08 (TGG ACC GGT G) GLC-10 (TGT CTG GGT G) GLC-11 (AAA GCT GCG G) GLC-12 (TGT CAT CCC C) GLC-13 (AAG CCT CGT C) GLC-14 (TGC GTG CTT G) GLC-15 (GAC GGA TCA G) GLC-16 (CAC ACT CCA G) GLC-18 (TGA GTG GGT G) GLC-19 (GTT GCC AGC C)		
M13 (GAG GGT GGC GGT TCT)	*Mesorhizobium* strains and *Phaseolus vulgaris*-nodulating isolates	[3, 4, 16]
OPAA10 (TGG TCG GGT G) OPACO3 (CAC TGG CCC A) OPACO4 (ACG GGG ACC TG) OPAE10 (AGC AGC GAG G)	*Rhizobium tropici* strains	[17]
OPB7 (GGT GAC GCA G) OPJ10 (AAG CCC GAG G)	*Ensifer meliloti* (formerly *Sinorhizobium meliloti*) strains	[18]
OPW-05 (CTG CTT CGA G) OPW-18 (GGC GCA ACT G) OPY-04 (AAG GCT CGA C) OPY-07 (GAC CGT CTG T)	*Ens. meliloti* strains and *Rhizobium sullae* strains	[19]
P1 (GTG TGT GTG TGT GTG TGT GT) P2 (GAC AGA CAG ACA GCA)	rhizobial strains	[20]
Primer-1 (GGT GCG GGA A) Primer-3 (AAG AGC CCG T)	*Bradyrhizobium* strains	[7, 21]
Primer-2 (GTT TCG CTC C)	*Bradyrhizobium* strains, *Derris elliptica*-nodulating isolates, *Indigofera tinctoria*-nodulating isolates and *Pueraria mirifica*-nodulating isolates	[2, 7, 21]
Primer-4 (AAG AGC CCG T) Primer-5 (AAC GCG CAA C) Primer-6 (CCC GTC AGC A)	*Bradyrhizobium* strains	[7]
PRIMM239 (CTG AAG CGG A)	*R. leguminosarum* strains	[14]
PucFor (GTA AAA CGA CGG CCA GT)	*R. leguminosarum* strains	[12]
RF2 (CGG CCC CTG T) 1247 (AAG AGC CCG T)	*Ens. meliloti* (formerly *Rhizobium meliloti* and *Sin. meliloti*) strains	[8, 18]

Table 1: cont…

RP04 (GGA AGT CGC C) RP05 (AGT CGT CCC C)	*Rhizobium* strains	[22]
SPH1 (GAC GAC GAC GAC GAC)	*R. leguminosarum* strains	[12]
879F (GCC TGG GGA GTA CGG CCG CA)	*Mesorhizobium* strains	[3]

5.2. TWO-PRIMERS RANDOM AMPLIFIED POLYMORPHIC DNA (TP-RAPD) ANALYSIS OF RHIZOBIA

Two-primers random amplified polymorphic DNA (TP-RAPD) is the peculiarity of RAPD in the use of pairs of larger primers (approximately 20 bp). TP-RAPD fingerprinting is rapid, sensitive, reliable, highly reproducible and suitable for a large number of microorganisms. TP-RAPD fingerprinting can be applied to bacterial taxonomy, ecological studies and for the detection of new bacterial species. TP-RAPD patterns are not strain dependent as the patterns of strains in the same species have been found to be identical [4, 23, 24]. No variations were observed in TP-RAPD patterns of strains in the same species with different plasmid profiles [23]. The TP-RAPD pattern of a novel rhizobial species, *Shinella kummerowiae,* was obviously different from the patterns of the 2 defined *Shinella* species and other rhizobial species. This supports the proposal of a novel species, *Shi. kummerowiae.* The TP-RAPD patterns of 3 *Shinella* species, 2 *Rhizobium* species, a species of each *Ensifer* and *Mesorhizobium* yielded 2 to 5 bands per lane in the sizes of less than 300 bp to 1.0 kb [25]. TP-RAPD was performed with *Herbaspirillum lusitanum* and *Ochrobactrum cytisi*, using the primer pair 879F (5' GCC TGG GGA GTA CGG CCG CA 3') and 1522R (5' AAG GAG GTG ATC CAN CC(A/G) CA 3'), and the primer pair 8F (5'AGA GTT TGA TCC TGG CTC AG 3') and 1522R, respectively [26, 27]. The PCR conditions were as follows: first denaturing at 95°C for 9 min; followed by 35 cycles of denaturing at 95°C for 1 min, annealing at 45°C for 1 min and extension at 72°C for 2 min; with final extension at 72°C for 7 min [28].

5.3. REPETITIVE SEQUENCE BASED PCR (rep-PCR) FINGERPRINTING OF RHIZOBIA

5.3.1. Principle of rep-PCR

Families of short intergenic repeated sequences have been found in enteric bacteria, primarily *Escherichia coli* and *Salmonella typhimurium* [29-32]. These sequences contain highly conserved central inverted repeats and can be divided in

2 classes that do not share significant homology. Class I consists of the repetitive extragenic palindromic (REP) elements [32] and Class II consists of the enterobacterial repetitive intergenic consensus (ERIC) sequences [30]. The actual functions of these highly repeated and conserved elements remain an enigma, although their involvement in stabilizing messenger RNA (mRNA), translational coupling between genes, homologous recombination, chromosome organization as well as binding HU protein (DNA binding protein), DNA gyrase and DNA polymerase I has been suggested [33].

The repetitive sequence based PCR (rep-PCR) fingerprints can be generated by primers derived from conserved repeat sequences present in bacterial genome [34]. These PCR based methods rely upon the same approach as RAPD, but different primers are used. These include pairs of primers for amplification of ERIC and REP sequences [12, 34] as well as a single primer for amplification of BOX sequence [30, 32, 35]. Refer to type of repetitive sequences, the methods are termed as "ERIC-PCR", "REP-PCR" and "BOX-PCR". The previous studies have demonstrated that rep-PCR methods are suitable for distinguishing strains at species or below levels [34-42].

5.3.2. Enterobacterial Repetitive Intergenic Consensus (ERIC)-PCR for Examining Genotypic Diversity of Rhizobia

A pair of primers and PCR cycling condition used for ERIC-PCR fingerprinting are shown in Table **2**. In general, the ERIC primer set generates multiple distinct DNA patterns including 1 to 15 fragments per lane, of sizes ranging from approximately 100 bp to 5.0 kb.

ERIC-PCR was applied to identify *Rhizobium* strains in environmental samples. Some strain groups from the non-polluted soil were suppressed in the polluted samples, and new strain groups were detected in the slurry-polluted soil, suggesting that one of the local parameters, slope position, had significantly greater impact on the composition of *Rhizobium* population than did the presence of slurry [43]. The ERIC primer set provided an effective mean for differentiating among *Mesorhizobium loti* (formerly *Rhizobium loti*) strains [44]. A high level of the genetic diversity among soybean rhizobia was detected by ERIC-PCR analysis [45]. Nodule occupancy with *Trifolium repens* L. (white clover) by 3 strains of

R. leguminosarum was studied by ERIC-PCR fingerprinting. The results showed that ERIC-PCR was an efficient and reliable method that can be used to identify *Rhizobium* strains directly from nodule suspensions [46]. ERIC-PCR was performed with rhizobial strains that nodulate *Phaseolus vulgaris* (common bean). The strains showing PCR-RFLP profiles similar to *Rhizobium etli* often occurred in quite dissimilar ERIC-PCR clusters, supporting the value of ERIC-PCR method for the evaluation in diversity of rhizobial strains, but not of rhizobial species. Although it is clearly that ERIC-PCR is less valuable for taxonomic purpose [47], this method has been found to be extremely sensitive and can detect minor differences among different strains of the same bacterial genus and species.

5.3.3. Repetitive Extragenic Palindromic (REP)-PCR for Examining Genotypic Diversity of Rhizobia

A pair of primers and PCR cycling condition used for REP-PCR fingerprinting are shown in Table **2**. In general, REP-PCR yields multiple distinct patterns with 1 to15 fragments, of sizes ranging from approximately 300 bp to 5.0 kb.

The genetic diversity of 44 rhizobial isolates from *Astragalus*, *Onobrychis* and *Oxytropis* spp. was evaluated by REP-PCR fingerprinting, revealing a high level of diversity within single 16S rDNA types [48]. The genetic diversity of salt-tolerant bacteria able to establish efficient symbiosis with *Lotus* spp. was performed by REP-PCR, revealing a relatively high level of genetic diversity among 120 isolates, with degrees of relatedness ranging from 20% to 97% [49].

5.3.4. BOX-PCR for Examining Genotypic Diversity of Rhizobia

A primer and PCR cycling condition used for BOX-PCR fingerprinting are shown in Table **2**. BOX-PCR fingerprint yields very complex patterns with approximately 30 fragments, of sizes ranging from approximately 200 bp to 5.0 kb. Cluster analysis can be performed using the Bionumerics software (Applied Mathematics, Belgium) based on the Pearson correlation method. Although the rep-PCR based methods are thought to be limited for the investigation of diversity of symbiotic genes among chromosomally closely related strains, Pongsilp *et al.* [50] indicated that the symbiotic and non-symbiotic strains of *Bradyrhizobium* could be separated into 2 distinct clusters based on rep-PCR fingerprinting, using the BOXA1R primer. Among 31 strains of *Rhizobium multihospitium* native of Xinjiang, China, 20 BOX patterns were generated and formed a cluster at 71% similarity [51]. BOX-PCR fingerprinting, together with phenotypic characterization, sodium dodecyl sulphate-polyacrylamide gel electrophoresis (SDS-PAGE) of whole-cell proteins and cellular fatty acid profiles, suggested the proposal of a novel species of legume symbionts, *Rhizobium vallis* [52]. BOX-PCR was employed to investigate the genetic diversity of *Mimosa*-nodulating rhizobia. The 30 distinct BOX-PCR genotypic patterns were detected, demonstrating the genetic and phenotypic diversity within 116 isolates of *Burkholderia* and *Cupriavidus* [53].

Table 2: Primer pairs used in rep-PCR fingerprintings

Type of PCR	Primer and Sequence (5' to 3')	PCR Mixture	PCR Cycling Condition	References
ERIC-PCR	Forward: ERIC2 (AAG TAA GTG ACT GGG GTG AGC G) Reverse: ERIC1R (ATG TAA GCT CCT GGG GAT TAC C)	2 ng/µl template DNA; 1.25 mM dNTPs; 2 µM each primer; 0.08 unit/µl *Taq* polymerase	95°C 7 min; 30 cycles x (94°C 1 min, 52°C 1 min and 65°C 8 min); 65°C 16 min	[34, 54]
REP-PCR	Forward: REP 2 (ICG ICT TAT CIG GCC TAC) Reverse: REP 1R (III ICG ICG ICA TCI GGC) (I is Inosine)	2 ng/µl template DNA; 1.25 mM dNTPs; 2 µM each primer; 0.08 unit/µl *Taq* polymerase	95°C 6 min; 30 cycles x (94°C 1 min, 40°C 1 min and 65°C 8 min); 65°C 16 min	[12, 34, 54]
BOX-PCR	BOXA1R: (CTA CGG CAA GGC GAC GCT GAC)	2 ng/µl template DNA; 1.25 mM dNTPs; 2 µM primer; 0.08 unit/µl *Taq* polymerase	95°C 2 min; 30 cycles x (94°C 3 sec, 92°C 30 sec, 50°C 1 min and 65°C 8 min)	[50, 55]

5.3.5. Advantages and Limitations of rep-PCR

The rep-PCR based methods have become valuable because of the following characteristics: i) these methods can be applied to both related and widely divergent strains by using the same set of primers; ii) these methods are adequate for differentiating closely related strains and/or determining the phylogenetic relationships; iii) the complexity of the patterns can be manipulated by the use of the different primers in various combinations; iv) simple agarose gel electrophoresis is sufficient to separate PCR products, making the procedures much less time-consuming, avoiding fastidious DNA extraction and hybridization; v) these methods offer convenient alternatives to conventional RFLP analysis with the same range of resolution levels and the same possibility of typing either whole genome or specific DNA regions; vi) these methods are suitable for large scale identification, classification of bacterial collections and study of large populations at intra-species level; vii) these methods generate highly reproducible fingerprints as compared to RAPD; viii) these methods are more discriminative than RAPD in many cases; ix) rep-PCR with BOX primer can reflect the variablility of symbiotic genes among rhizobial populations; x) a dendrogram can be constructed from the rep-PCR patterns to present the genetic relatedness among rhizobial populations. However, like RAPD, the limitation of rep-PCR based methods is that they are less valuable for taxonomic purpose.

5.4. COMPARISONS OF RESOLUTION LEVEL AND CONSISTENCY AMONG RAPD, ERIC-, REP- AND BOX-PCR

The methods, including ERIC-PCR, REP-PCR and BOX-PCR, have been used to type several rhizobial strains and offer alternatives or additional approaches for the measurement of the genetic diversity within rhizobial species [12, 13]. de Bruijn [36] demonstrated the usefulness of DNA fingerprinting by PCR using ERIC and REP primers for the identification and the classification of members of *Azorhizobium*, *Bradyrhizobium* and *Rhizobium* species. Vinuesa *et al.* [56] reported the use of rep-PCR fingerprinting with ERIC, REP and BOX primers to study the genotypic diversity among *Bradyrhizobium* strains and exploited the taxonomic resolution of rep-PCR by combining ERIC-, REP- and BOX-PCR genomic fingerprints, maximizing strain discrimination and the phylogenetic coherency of the obtained cluster. ERIC- and REP-PCR analyses revealed a high level of the genetic diversity of soybean rhizobia in Paraguay, with the majority of the isolates representing unique strains [45]. Gao *et al.* [57] investigated the genetic diversity among 95 isolates from *Astragalus adsurgens* and found that all of the isolates and 24 reference strains could be differentiated by ERIC-, REP- and BOX-PCR fingerprinting analyses. The reproducible and specific polymorphic banding patterns were obtained with ERIC and REP primers for *Ens.*

meliloti (formerly *Sinorhizobium meliloti*), whereas BOX primer did not reveal any polymorphism [19]. The high intra-species diversity of *R. leguminosarum* strains isolated from *Vicia sativa* (wild vetch) was observed in ERIC-, REP- and BOX-PCR fingerprintings [58].

The comparisons of resolution level, consistency and reproducibility obtained from RAPD and rep-PCR based methods have been demonstrated. Groupings of 43 strains of *R. leguminosarum* biovars viciae, trifolii and phaseoli, generated by PCR DNA fingerprinting with either REP primer or 2 different RAPD primers, were correlated with similar resolution levels [12]. Laguerre *et al.* [48] reported that both ERIC- and REP-PCR analyses yielded similar resolution levels. Dendrograms derived from RAPD, ERIC and REP profiles of indigenous soybean rhizobia were consistent with one another [59]. The DNA fingerprints observed by ERIC and REP primers were highly reproducible as compared to RAPD [19]. ERIC-PCR fingerprinting was found more discriminative among 14 isolates of *Mes. loti* than did RAPD [44]. ERIC- and REP-PCR analyses were applied to the identification and the classification of 44 strains of *Bradyrhizobium japonicum*, 7 strains of *Ens. meliloti* and 10 strains of *R. leguminosarum* native to Japan and Thailand. Both ERIC and REP primers induced reproducible PCR band patterns, although REP-PCR generated more bands and appeared to be more useful for distinguishing the isolates from each other [60]. The rep-PCR with ERIC and REP primers were found to be more efficient than RAPD for genotyping *Ens. meliloti* and *Rhizobium sullae* isolates [19]. ERIC- and BOX-PCR showed very similar, almost identical, grouping of strains which demonstrated that these methods are reliable and suitable for rhizobial strain identification. In general, the fingerprints which were generated with BOX primer, produced the highest number of polymorphic bands. Therefore, the BOX-PCR fingerprints showed the highest genetic polymorphism when compared with ERIC- and REP-fingerprints [58]. The previous studies showed that ERIC- and REP-PCR reflected the variability of chromosomal DNA regions but did not reflect the variability of symbiotic gene regions in *R. leguminosarum* species and biovars [12], while BOX-PCR could separate the symbiotic strains and non-symbiotic strains of *Bradyrhizobium* into 2 distinct clusters [50].

5.5. AMPLIFIED FRAGMENT LENGTH POLYMORPHISM (AFLP) FINGERPRINTING OF RHIZOBIA

5.5.1. Principle of AFLP

Amplified fragment length polymorphism (AFLP) is one of the PCR based DNA fingerprinting techniques. AFLP overcomes problems of RAPD and RFLP. The AFLP technique is a highly discriminating fingerprinting method, based on the selective PCR amplification of certain restriction fragments from a digestion of total genomic DNA. In AFLP analysis, there are 3 major steps in the procedure: i) digestion of total genomic DNA with 2 endonucleases and ligation of double-stranded oligonucleotide adapters. These consist of a core sequence and an enzyme-specific sequence served as primer sites for amplification of the restriction products; ii) selective amplification of the restriction fragments by PCR using primer pairs containing common sequences of the adapter and 1 to 3 arbitrary nucleotides; iii) analysis of the amplified fragments using polyacrylamide gel electrophoresis [61, 62]. In the procedure, 1 µg of total genomic DNA is digested by 2 endonucleases. Double-stranded oligonucleotide adapters with a single-stranded overhang homologous to the 5' and 3' ends generated during restriction are ligated to the DNA fragments. The ligated DNA fragments are amplified by PCR using primers complementary to the adapter and restriction site sequence with additional selective nucleotides at the 3' end. The amplified products are separated by polyacrylamide gel electrophoresis. The cluster analysis can be performed from AFLP patterns using the UPGMA clustering algorithm. The most suitable enzyme combination is the one that produces multiple fragments (30 to 50) in different lengths resulting in an evenly distributed banding pattern. For *Bradyrhizobium* species which have a high GC percentage in their genomes, the combination of *Taq*I (restriction site: T/CGA) and *Apa*I (restriction site: GGGCC/C) is recommended [63]. The combination of different restriction enzymes and the choice of selective nucleotides in the primers for PCR make AFLP a useful new system for molecular typing of microorganisms [61]. AFLP has been reported to be suitable for distinguishing strains at the species and below levels [34-42]. Janssen *et al.* [38] demonstrated the superior discriminative power of AFLP toward the differentiation of highly related bacterial strains that belong to the same species or even biovar (intra-subspecific level). For investigating the whole genome, AFLP has been reported as the most discriminative technique [63].

5.5.2. AFLP for Examining Genotypic Diversity of Rhizobia

The diversity of bradyrhizobia from *Faidherbia albida* and various *Aeschynomene* species was estimated using AFLP analysis. The AFLP technique was shown to provide an insight into the extent of genotypic diversity of *Bradyrhizobium* isolates. As a grouping method for new isolates, it was superior to amplified 16S rDNA restriction analysis (ARDRA), analysis of metabolic profiles using the Phenotype Micro Array (PM) system (Biolog) (Biolog Inc., CA) and SDS-PAGE analysis of proteins because of its higher taxonomic resolution [64]. The 64 *Bradyrhizobium* strains isolated from nodules of 27 native leguminous plant species in Senegal were characterized into 27 groups by using AFLP [63]. Gao *et al.* [57] used molecular biological methods to investigate the genetic diversity among 95 isolates from *Ast. adsurgens.* All isolates and 24 reference strains could be differentiated by AFLP, REP-, ERIC- and BOX-PCR fingerprinting analyses and some of the AFLP groups also covered several rep-PCR groups. AFLP data produced grouping in line with analysis of the sequences of 16S-23S rRNA intergenic spacer (IGS) region data. *Bradyrhizobium* strains belonging to the same AFLP group were found to belong to the same genospecies [65]. The grouping of *Bradyrhizobium* strains by AFLP was consistent with IGS PCR-RFLP and ARDRA analyses. The AFLP analysis, which is based on whole-genome variability, showed that the Caucasian *Rhizobium galegae* biovar orientalis populations were more diverse than *R. galegae* biovar officinalis populations [66].

5.5.3. Advantages and Limitations of AFLP

AFLP overcomes problems of RAPD and RFLP with these benefits: i) it is a highly discriminating fingerprinting method as it can differentiate at species or below levels; ii) it has been reported as the most discriminative techniques for invesrigating the whole genome as compared to other molecular techniques such as ARDRA and IGS PCR-RFLP; iii) the results have been reported to be consistent with those obtained by DNA-DNA hybridization, ARDRA and IGS PCR-RFLP; iv) it has a similar discriminating power as DNA-DNA hybridization but less time- and labor-consuming; v) it is a potentially useful for taxonomic proposes. However, the limitation of AFLP has been previously reported. The AFLP procedure is rather laborious and it could not reflect more remote relationships (DNA homology level of 40% to 60%) between species [64, 65].

REFERENCES

[1] Rafalski JA, Tingey SV, William JGK. RAPD markers-a new technology for genetic mapping and plant breeding. AgBiotech News Inform 1991; 3: 645-48.

[2] Pongsilp N, Nuntagij A. Genetic diversity and metabolites production of root-nodule bacteria isolated from medicinal legumes *Indigofera tinctoria*, *Pueraria mirifica* and *Derris elliptica* Benth. grown in different geographic origins across Thailand. Amer-Eur J Agric Envir Sci 2009; 6: 26-34.

[3] Rivas R, Peix A, Mateos PF, *et al.* Biodiversity of populations of phosphate solubilizing rhizobia that nodulates chickpea in different Spanish soils. In: Velazquez E, Rodriguez-Barrueco C, Eds. First International Meeting on Microbial Phosphate Solubilization. Dordrecht, Springer, 2007; pp. 23-33.

[4] Valverde A, Igual JM, Peix A, *et al. Rhizobium lusitanum* sp. nov. a bacterium that nodulates *Phaseolus vulgaris.* Int J Syst Evol Microbiol 2006; 56: 2631-37.

[5] Lunge VR, Ikuta N, Fonseca ASK, *et al.* Identification and inter relationship analysis of *Bradyrhizobium japonicum* strains by restriction fragment length polymorphism (RFLP) and random amplified polymorphic DNA (RAPD). World J Microbiol Biotechnol 1994; 10: 648-52.

[6] Nishi CYM, Boddey LH, Vargas MAT, Hungria M. Morphological, physiological and genetic characterization of two new *Bradyrhizobium* strains recently recommended as Brazilian comercial inoculants for soybean. Symbiosis 1996; 20: 147-62.

[7] Nuntagij A, Abe M, Uchiumi T, *et al.* Characterization of *Bradyrhizobium* strains isolated from soybean cultivation in Thailand. J Gen Appl Microbiol 1997; 43: 183-87.

[8] Paffetti D, Scotti C, Gnocchi S, *et al.* 1996. Genetic diversity of an Italian *Rhizobium meliloti* population from different *Medicago sativa* varieties. Appl Envir Microbiol 62: 2279-85.

[9] Niemann S, Puhler A, Tichy HV, *et al.* Evaluation of the resolving power of three different DNA fingerprinting methods to discriminate among isolates of a natural *Rhizobium meliloti* population. J Appl Microbiol 1997; 82: 477-84.

[10] de Oliveira IR, Vasconcellos MJ, Seldin L, *et al.* Random amplified polymorphic DNA analysis of effective *Rhizobium* sp. associated with beans cultivated in Brazil Cerrado soils. Braz J Microbiol 2000; 31: 39-44.

[11] Valverde A, Velazquez E, Cervantes E, *et al.* Evidence of an American origin for symbiosis-related genes in *Rhizobium lusitanum.* Appl Envir Microbiol 2011; 77: 5665-70.

[12] Laguerre G, Mavingui P, Allard MR, *et al.* Typing of rhizobia by PCR DNA fingerprinting and PCR-restriction fragment length polymorphism analysis of chromosomal and symbiotic gene regions: application to *Rhizobium leguminosarum* and its different biovars. Appl Envir Microbiol 1996; 62: 2029-36.

[13] van Berkum P, Eardly BD. Molecular Evolutionary Systematics of the Rhizobiaceae. In: Spaink HP, Kondorosi A, Hooykaas PJJ, Eds. The Rhizobiaceae: Molecular Biology of Model Plant-Associated Bacteria. Dordrecht, Kluwer Academic Publishers, 1998; pp. 1-24.

[14] Moschetti G, Peluso A, Protopapa A, *et al.* Use of nodulation pattern, stress tolerance, *nodC* gene amplification, RAPD-PCR and RFLP-16S rDNA analysis to discriminate genotypes of *Rhizobium leguminosarum* biovar *viciae.* Syst Appl Microbiol 2005; 28: 619-31.

[15] Sajjad M, Malik TA, Arshad M, *et al.* PCR studies on genetic diversity of rhizobial strains. Int J Agric Biol 2008; 10: 505-10.

[16]   Zurdo-Pineiro JL, Garcia-Fraile P, Rivas R, *et al.* Rhizobia from Lanzarote, the Canary Islands, that nodulate *Phaseolus vulgaris* have characteristics in common with *Sinorhizobium meliloti* isolates from mainland Spain. Appl Envir Microbiol 2009; 75: 2354-59.

[17]   Raposeiras R, Marriel IE, Muzzi MRC, *et al. Rhizobium* strains competitiveness on bean nodulation in Cerrado soils. Pesq Agropec Bras 2006; 41: 439-47.

[18]   Carelli M, Gnocchi S, Fancelli S, *et al.* Genetic diversity and dynamics of *Sinorhizobium meliloti* populations nodulating different alfalfa cultivars in Italian soils. Appl Envir Microbiol 2000; 66: 4785-89.

[19]   Elboutahiri N, Thami-Alami I, Zaid E, Udupa SM. Genotypic characterization of indigenous *Sinorhizobium meliloti* and *Rhizobium sullae* by rep-PCR, RAPD and ARDRA analyses. Afr J Biotechnol 2009; 8: 979-85.

[20]   El-Fiki AA. Genetic diversity in rhizobia determined by random amplified polymorphic DNA analysis. J Agric Soc Sci 2006; 2: 1-4.

[21]   Pongsilp N, Nuntagij A. Selection and characterization of mungbean root nodule bacteria based on their growth and symbiotic ability in alkaline conditions. Suranaree J Sci Technol 2007; 14: 277-86.

[22]   Richardson AE, Viccars LA, Watson JM, Gibson AH. Differentiation of *Rhizobium* strains using the polymerase chain reaction with random and directed primers. Soil Biol Biochem 1995; 27: 515-24.

[23]   Rivas R, Velazquez E, Valverde A, *et al.* A two primers random amplified polymorphic DNA procedure to obtain polymerase chain reaction fingerprints of bacterial species. Electrophoresis 2001; 22: 1086-89.

[24]   Rivas R, Abril A, Trujillo ME, Velazquez E. *Sphingomonas phyllosphaerae* sp. nov., from the phyllosphere of *Acacia caven* in Argentina. Int J Syst Evol Microbiol 2004; 54: 2147-50.

[25]   Lin DX, Wang ET, Tang H, *et al. Shinella kummerowiae* sp. nov., a symbiotic bacterium isolated from root nodules of the herbal legume *Kummerowia stipulacea*. Int J Syst Evol Microbiol 2008; 58: 1409-13.

[26]   Valverde A, Velazquez E, Gutierrez C, *et al. Herbaspirillum lusitanum* sp. nov., a novel nitrogen-fixing bacterium associated with root nodules of *Phaseolus vulgaris*. Int J Syst Evol Microbiol 2003; 53: 1979-83.

[27]   Zurdo-Pineiro JL, Rivas R, Trujillo ME, *et al. Ochrobactrum cytisi* sp. nov., isolated from nodules of *Cytisus scoparius* in Spain. Int J Syst Evol Microbiol 2007; 57: 784-88.

[28]   Trujillo ME, Willems A, Abril A, *et al.* Nodulation of *Lupinus albus* by strains of *Ochrobactrumlupini* sp. nov. Appl Envir Microbiol 2005; 71: 1318-27.

[29]   Gilson E, Clement JM, Brutlag D, Hofnung M. A family of dispersed repetitive extragenic palindromic DNA sequences in *E. coli*. EMBO J 1984; 3: 1417-21.

[30]   Hulton CSJ, Higgins CF, Sharp PM. ERIC sequences: a novel family of repetitive elements in the genomes of *Escherichia coli, Salmonella typhimurium* and other enteric bacteria. Mol Microbiol 1991; 5: 825-34.

[31]   Sharples GJ, Lloyd RG. A novel repeated DNA sequence located in the intergenic regions of bacterial chromosomes. Nucleic Acids Res 1990; 18: 6503-08.

[32]   Stern MJ, Ames GFL, Smith NH, *et al.* Repetitive extragenic palindromic sequences: a major component of the bacterial genome. Cell 1984; 37: 1015-26.

[33]   Rashid MH, Sattar MA, Uddin MI, Young JPW. Molecular characterization of symbiotic root nodulating rhizobia isolated from lentil (*Lens culinaris* Medik.). EJEAFChe 2009; 8: 602-12.

[34]   Versalovic J, Koeuth T, Lupski JR. Distribution of repetitive DNA sequences in eubacteria and application to fingerprinting of bacterial genomes. Nucleic Acids Res 1991; 19: 6823-31.

[35]   Martin R, Humbert O, Camara M, *et al.* A highly conserved repeated DNA element located in the chromosome of *Streptococcus pneumoniae.* Nucleic Acids Res 1992; 20: 3479-83.

[36]   de Bruijn FJ. Use of repetitive (repetitive extragenic palindromic and enterobacterial repetitive intergeneric consensus) sequences and the polymerase chain reaction to fingerprint the genomes of *Rhizobium meliloti* isolates and other soil bacteria. Appl Envir Microbiol 1992; 58: 2180-87.

[37]   Huys G, Coopman R, Janssen P, Kersters K. High-resolution genotypic analysis of the genus *Aeromonas* by AFLP fingerprinting. Int J Syst Bacteriol 1996; 46: 572-80.

[38]   Janssen P, Coopman R, Huys G, *et al.* Evaluation of the DNA fingerprinting method AFLP as new tool in bacterial taxonomy. Microbiol 1996; 142: 1881-93.

[39]   Nick G, Lindstrom K. Use of repetitive sequences and the polymerase chain reaction to fingerprint the genomic DNA of *Rhizobium galegae* strains and to identify the DNA obtained by sonicating the liquid cultures and root nodules. Syst Appl Microbiol 1994; 17: 265-73.

[40]   Nick G, de Lajudie P, Eardly BD, *et al. Sinorhizobium arboris* sp. nov. and *Sinorhizobium kostiense* sp. nov., isolated from leguminous trees in Sudan and Kenya. Int J Syst Bacteriol 1999; 49: 1359-68.

[41]   Versalovic J, Schneider M, de Bruijn FJ, Lupski LR. Genomic fingerprinting of bacteria using repetitive sequence-based polymerase chain reaction. Methods Mol Cell Biol 1994; 5: 25-40.

[42]   Vos P, Hogers R, Bleeker M, *et al.* AFLP: a new technique for DNA fingerprinting. Nucleic Acids Res 1995; 23: 4407-14.

[43]   Labes G, Ulrich A, Lentzsch P. Influence of bovine slurry deposition on the structure of nodulating *Rhizobium leguminosarum* bv. viciae soil populations in a natural habitat. Appl Envir Microbiol 1996; 62: 1717-22.

[44]   Agius F, Sanguinetti C, Monza J. Strain-specific fingerprints of *Rhizobium loti* generated by PCR with arbitrary and repetitive sequences. FEMS Microbiol Ecol 1997; 24: 87-92.

[45]   Chen LS, Figueredo A, Pedrosa FO, Hungria M. Genetic characterization of soybean rhizobia in Paraguay. Appl Envir Microbiol 2000; 66: 5099-03.

[46]   Svenning MM, Gudmundssont J, Fagerli IL, Leinonen P. Competition for nodule occupancy between introduced strains of *Rhizobium leguinnosarum* biovar *trifolii* and its influence on plant production. Ann Bot 2001; 88: 781-87.

[47]   Grange L, Hungria M. Genetic diversity of indigenous common bean (*Phaseolus vulgaris*) rhizobia in two Brazilian ecosystems. Soil Biol Biochem 2004; 36: 1389-98.

[48]   Laguerre G, van Berkum P, Amarger N, Prevost D. Genetic diversity of rhizobial symbionts isolated from legume species within the genera *Astragalus*, *Oxytropis*, and *Onobrychis*. Appl Envir Microbiol 1997; 63: 4748-58.

[49]   Lorite MJ, Munoz S, Olivares J, *et al.* Characterization of strains unlike *Mesorhizobium loti* that nodulate *Lotus* spp. in saline soils of Granada, Spain. Appl Envir Microbiol 2010; 76: 4019-26.

[50]   Pongsilp N, Teaumroong N, Nuntagij A, *et al.* Genetic structure of indigenous non-nodulating and nodulating populations of *Bradyrhizobium* in soils from Thailand. Symbiosis 2002; 33: 39-58.

[51]   Han TX, Wang ET, Wu LJ, *et al. Rhizobium multihospitium* sp. nov., isolated from multiple legume species native of Xinjiang, China. Int J Syst Evol Microbiol 2008; 58: 1693-99.

[52]   Wang F, Wang ET, Wu LJ, *et al. Rhizobium vallis* sp. nov., isolated from nodules of three leguminous species. Int J Syst Evol Microbiol 2010; 61: 2582-88.

[53]   Liu XY, Wu W, Wang ET, *et al.* Phylogenetic relationships and diversity of β-rhizobia associated with *Mimosa* species grown in Sishuangbanna, China. Int J Syst Evol Microbiol 2011; 61: 334-42.

[54]   Ogutcu H, Adiguzel M, Gulluce M, *et al.* Molecular characterization of *Rhizobium* strains isolated from wild chickpeas collected from high altitudes in Erzurum-Turkey. Rom Biotechnol Lett 2009; 14: 4294-99.

[55]   Versalovic J, de Bruijn FJ, Lupski JR. Repetitive Sequence Based PCR (rep-PCR) DNA Fingerprinting of Bacterial Genomes. In: de Bruijn FJ, Lupski JR, Weinstock GM, Eds. Bacterial Genomes: Physical Structure and Analysis. New York, Chapman and Hall, 1998; pp. 437-54.

[56]   Vinuesa P, Rademaker JLW, de Bruijn FJ, Werner D. Genotypic characterization of *Bradyrhizobium* strains nodulating endemic woody legumes of the Canary islands by PCR-restriction fragment length polymorphism analysis of genes encoding 16S rRNA (16S rDNA) and 16S-23S rDNA intergenic spacers, repetitive extragenic palindromic PCR genomic fingerprinting, and partial 16S rDNA sequencing. Appl Envir Microbiol 1998; 64: 2096-04.

[57]   Gao JL, Terefework ZD, Chen WX, Lindstrom K. Genetic diversity of rhizobia isolated from *Astragalus adsurgens* growing in different geographical regions of China. J Biotechnol 2001; 91: 155-68.

[58]   Adiguzel A, Ogutcu H, Baris O, *et al.* Isolation and characterization of *Rhizobium* strains from wild vetch collected from high altitudes in Erzurum-Turkey. Rom Biotechnol Lett 2010; 15: 5017-24.

[59]   Sikora S, Redzepovic S. Genotypic characterisation of indigenous soybean rhizobia by PCR-RFLP of 16S rDNA, rep-PCR and RAPD Analysis. Food Technol Biotechnol 2003; 41: 61-67.

[60]   Tajima S, Hirashita T, Yoshihara K, *et al.* Application of repetitive extragenic palindromic (REP)-PCR and enterobacterial repetitive intergenic consensus (ERIC)-PCR analysis to the identification and classification of Japan and Thai local isolates of *Bradyrhizobium japonicum, Sinorhizobium meliloti*, and *Rhizobium leguminosarum*. Soil Sci Plant Nutr 2000; 46: 241-47.

[61]   Lin JJ, Kuo J, Ma J. A PCR-based DNA fingerprinting technique: AFLP for molecular typing of bacteria. Nucleic Acids Res 1996; 24: 3649-50.

[62]   Strange RH. Introduction to Plant Pathology. Wiley, New York, USA. 2003.

[63]   Doignon-Bourcier F, Willems A, Coopman R, *et al.* Genotypic characterization of *Bradyrhizobium* strains nodulating small Senegalese legumes by 16S-23S rRNA intergenic gene spacers and amplified fragment length polymorphism fingerprint analyses. Appl Envir Microbiol 2000; 66: 3987-97.

[64]   Willems A, Doignon-Bourcier F, Coopman R, *et al.* AFLP fingerprint analysis of *Bradyrhizobium* strains isolated from *Faidherbia albida* and *Aeschynomene* species. Syst Appl Microbiol 2000; 23: 137-47.

[65]   Willems A, Coopman R, Gillis M. Comparison of sequence analysis of 16S-23S rDNA spacer regions, AFLP analysis and DNA-DNA hybridizations in *Bradyrhizobium*. Int J Syst Evol Microbiol 2001; 51: 623-32.

[66]   Andronov EE, Terefework Z, Roumiantseva ML, *et al.* Symbiotic and genetic diversity of *Rhizobium galegae* isolates collected from the *Galega orientalis* gene center in the Caucasus. Appl Envir Microbiol 1995; 69: 1067-74.

CHAPTER 6

Genotypic Diversity of Rhizobia Assessed by Polymerase Chain Reaction-Restriction Fragment Length Polymorphism (PCR-RFLP)

Neelawan Pongsilp[*]

Department of Microbiology, Faculty of Science, Silpakorn University, Nakhon Pathom, Thailand

Abstract: Polymerase chain reaction-restriction fragment length polymorphism (PCR-RFLP) is used in determining the genetic relationships based upon PCR and restriction analysis. Specific genes, such as small subunit ribosomal RNA gene (16S rDNA), large subunit ribosomal RNA gene (23S rDNA), 16S-23S rRNA intergenic spacer (IGS) and symbiotic genes, have been used in PCR-RFLP. The PCR-RFLP profile is used to estimate the genetic diversity of microorganisms. The PCR-RFLP method has been used successfully in the differentiation of rhizobial species.

Keywords: Amplified 16S rDNA restriction analysis (ARDRA), Dendrogram, Endonuclease, Genotypic diversity, Large subunit ribosomal RNA gene (23S rDNA), Polymerase chain reaction-restriction fragment length polymorphism (PCR-RFLP), Rhizobia, Small subunit ribosomal RNA gene (16S rDNA), Symbiotic gene, 16S-23S rRNA intergenic spacer (IGS).

6.1. PRINCIPLE OF POLYMERASE CHAIN REACTION-RESTRICTION FRAGMENT LENGTH POLYMORPHIISM (PCR-RFLP)

Polymerase chain reaction-restriction fragment length polymorphism (PCR-RFLP) is used in determining the genetic relationships based upon PCR and restriction analysis. Specific regions of the genome are amplified and fingerprint patterns are obtained after restriction digestion of the amplification products. The PCR product suitable for PCR-RFLP should be relatively long to ensure maximum specificity of RFLP fingerprints. The 4-base-cutting endonucleases (restriction enzymes) are preferable to provide more chance of digestion. The PCR product is digested individually with several (generally, 2 to 9)

*Address correspondence to Neelawan Pongsilp: Department of Microbiology, Faculty of Science, Silpakorn University, Nakhon Pathom, Thailand; Tel: +66-34-245337; Fax: +66-34-245336-37; E-mail: neelawan@su.ac.th

endonucleases. The frequently used endonucleases are *Alu*I, *Cfo*I, *Dde*I, *Hae*III, *Hha*I, *Hinf*I, *Msp*I, *Nde*II, *Rsa*I, *Sau*3A1 and *Taq*I. Genes and endonucleases frequently used in PCR-RFLP are listed in Table **1**. Restriction fragments are separated by electrophoresis in 3% agarose gel. Restriction fragments shorter than 100 base pairs (bp) are not well resolved by electrophoresis, thus these fragments are not used in the analysis. The PCR-RFLP profile is obtained from the combination of the banding pattern across different enzyme digests and is used to estimate the genetic diversity of microorganisms. The PCR-RFLP profiles are transformed to genetic distances and a dendrograms is constructed from the data using the unweighted pair group method with arithmetic mean (UPGMA) or the neighbor-joining (NJ) tree-building method. The PCR-RFLP method has been used successfully in the differentiation of *Rhizobium* species [1, 2]. It provides a rapid tool for the identification of rhizobia and the detection of new taxa. Data also proves the usefulness of the PCR-RFLP method for rapidly identifying the known relatives of unclassified strains or new isolates [1].

6.2. PCR-RFLP OF SMALL SUBUNIT RIBOSOMAL RNA GENE (16S rDNA) OF RHIZOBIA

The gene encoding small subunit ribosomal RNA, termed as "16S rRNA gene" or "16S rDNA", has turned out to be a very good tool for the assessment of organismal phylogenies down to the genus level [3]. In 16S PCR-RFLP (also the same as "amplified 16S rDNA restriction analysis (ARDRA)"), pairs of universal primers, such as fD1 and rD1 [4] as well as Y1 and Y3 [5], are used for PCR amplification of the nearly full-length 16S rDNA (approximately 1.5 kilobases (kb)). Sequence divergences between 16S rDNA regions of pairs of strains can be estimated from the proportion of shared restriction fragments by the Nei and Li method [6].

The 16S PCR-RFLP has been used to determine the phylogenic positions of rhizobia including *Ensifer meliloti* (formerly *Sinorhizobium meliloti*) [7], *Mesorhizobium* [8] and *Rhizobium* [3, 7, 9, 10]. It has been reported to be useful in rhizobial taxonomy since the results are in good agreement with those from 16S rDNA sequence analysis, DNA-DNA hybridization, multilocus enzyme electrophoresis (MLEE) and numerical taxonomy [1, 3, 11-13]. It has become a rapid tool for the differentiation and the estimation of the genetic relationships among rhizobia at species and higher levels [1].

The 104 isolates of rhizobial symbionts of *Sesbania herbacea*, *Rhizobium galegae* type strain and *Rhizobium huautlense* type strain, had similar, but distinguishable, 16S rDNA as revealed by PCR-RFLP analysis [13]. The 55 isolates from root nodules of *Amorpha fruticosa* grouped by PCR-RFLP of 16S rDNA were also separated into groups by variation in MLEE profiles and DNA-DNA hybridization [8]. The identical 16S PCR-RFLP patterns were found among the 50 isolates from *Mimosa affinis* and *Rhizobium etli* type strain in individual digestions with 4 endonucleases. This result indicated a close phylogenetic relationship between the isolates and *R. etli* type strain [10]. The 48 rhizobial isolates from root nodules of *Indigofera* sp. and *Kummerowia* sp. were studied by 16S PCR-RFLP. In a cluster analysis of the RFLP patterns, cluster 1 was grouped with *Rhizobium* species and cluster 2 fell in the genus *Sinorhizobium* [14]. In 16S PCR-RFLP of rhizobia nodulating *Phaseolus vulgaris* (common bean), strains showing PCR-RFLP profiles similar to *R. etli* often occurred in quite dissimilar enterobacterial repetitive intergenic consensus (ERIC)-PCR clusters [5]. The strains nodulating *Vicia faba* (faba bean) were characterized based on PCR-RFLP of both 16S rDNA and large subunit ribosomal RNA gene (23S rDNA). The strains could be divided into 6 genotypes based on 16S rDNA and 5 genotypes based on 23S rDNA [15].

The PCR-RFLP of 16S rDNA is a fast and simple technique that has been widely used for the genotypic characterization [1]. In some case, it can be performed with less number of enzymes to achieve the maximum discriminating power. The 16S PCR-RFLP with 9 endonucleases (*Alu*I, *Dde*I, *Hae*III, *Hha*I, *Hin*fI, *Msp*I, *Nde*II, *Rsa*I and *Taq*I) had no greater discriminating power than a combination of only 4 enzymes (*Hha*I, *Hin*fI, *Msp*I and *Rsa*I) [16].

Besides 16S rDNA, some other genes have been used in PCR-RFLP. The 23S rDNA and 16S-23S rRNA intergenic spacer (IGS) region have been used as markers of the genomic background of bacteria [3, 9, 17, 18].

6.3. PCR-RFLP OF LARGE SUBUNIT RIBOSOMAL RNA GENE (23S rDNA) OF RHIZOBIA

The 23S rRNA gene (23S rDNA) encodes large subunit ribosomal RNA. Primer 3 (5' CCG TGA GGG AAA GGT GAA AAG TACC 3'), which corresponds to the sequence from 461 to 485 bp in *Escherichia coli* nomenclature, and primer 4 (5' CCC GCT TAG ATG CTT TCA GC 3'), which corresponds to sequence from 2,744 to 2,763 bp in *Esc. coli*, can be used for the amplification of 23S rDNA [3]. Among the 78 strains isolated from root nodules of legumes growing in southern Ethiopia and 25 reference strains, PCR-RFLP of 23S rDNA delineated 103 different 23S rRNA genotypes. It was found that UPGMA dendrograms generated from cluster analyses of PCR-RFLP of 16S and 23S rDNA were in good agreement [19]. In some cases, the 23S dendrogram showed deeper branching than the 16S dendrogram and more genotypes were resolved, although the sequence divergence is not particularly high. The RFLP analysis of 42 rhizobial and agrobacterial strains resulted in 27 and 32 different restriction patterns for 16S and 23S rDNA, respectively, which *R. galegae* was phylogenetically distinct from other rhizobia and agrobacteria [3].

6.4. PCR-RFLP OF 16S-23S rDNA INTERGENIC SPACER (IGS) REGION OF RHIZOBIA

The intergenic spacer (IGS) between 16S and 23S rRNA sequences, termed as "16S-23S rRNA intergenic spacer (IGS) region", has been employed as a powerful tool to resolve the genetic variation among bacterial isolates. This region contains greater variability than 16S rDNA and is suitable in order to examine chromosomally encoded genetic variations at the intra-species level. For IGS PCR-RFLP, the primer pairs used as universal primers specific to IGS regions were designed. The primer pair FGPS1490 (5' TGC GGC TGG ATC ACC TCC TT 3') (corresponds to 3' end of 16S rDNA) and FGPS132 (5' CCG GGT TTC CCCATT CGG 3') (corresponds to 5' end of 23S rDNA) can be used to amplify IGS region. The PCR cycling condition is as follows: first denaturing at 95°C for 3 min; 35 cycles of denaturing at 94°C 1 min, annealing at 55°C for 1 min and extension at 72°C for 2 min; final extension at 72°C for 3 min. Length variability of IGS regions in both inter- and intra-species of rhizobia has been reported, as shown in Table **2**. Length variability of IGS regions could be in part explained by the presence of several tRNA genes varying in number and type in IGS regions. The discriminating power was sufficient to group chromosomally closely related strains on the basis of the simple, reproducible, and hence, easily analyzable patterns of restriction fragments [17]. The discrimination of isolates obtained from IGS polymorphism corresponded reasonably well to the rhizobial classification [20].

By IGS PCR-RFLP using 8 endonucleases, the 104 strains of *Bradyrhizobium* produced 70 types of combined restriction profiles, forming 16 groups [21]. The diversity of background genotypes of *Rhizobium leguminosarum* biovar viciae populations nodulating *Pisum sativum* (pea), *V. faba* and *Vicia sativa* (vetch) was examined by IGS PCR-RFLP. The diversity was higher in the vetch population than in the faba bean and pea populations, in which 16, 5 and 1 genotypes were detected, respectively. The results obtained from IGS-RFLP with 3 endonucleases supports the proposal of a novel species, *Rhizobium miluonense*, as it differed from those of the related *Rhizobium* species including *R. lusitanum*, *R. rhizogenes* (currently *Agrobacterium rhizogenes*) and *R. tropici*, with similarities less than 81% [22].

6.5. PCR-RFLP OF 16S rDNA-IGS OF RHIZOBIA

Instead of using either 16S rDNA or IGS alone, the region containing both 16S rDNA and IGS, also termed as "16S-IGS", has been used for PCR-RFLP. PCR-RFLP of 16S-IGS is useful for distinguishing among rhizobial genera and species and can be used for a rapid tracking of the known relatives of new isolates [23]. The primer pair FGPS6-63 (5' GGA GAG TTA GAT CTT GGC TCA G 3') (corresponds to 5' end of 16S rDNA) and FGPL132'-38 (5' CCG GGT TTC CCC ATT CGG 3') (corresponds to 5' end of 23S rDNA) can be used to ampify the region containing 16S rDNA and IGS. The PCR cycling condition is as follows: first denaturing at 95°C for 3 min; 35 cycles of denaturing at 95°C for 1 min, annealing at 55°C for 1 min and extension at 72°C for 2 min; final extension at 72°C for 3 min [24, 25].

The 30 rhizobial strains that nodulate *Cicer arietinum* L. (chickpea) were characterized by 16S-IGS PCR-RFLP. The 7 endonucleases generated 4 to 12 fragments per profile and a total of 178 different bands. The clusters obtained from PCR-RFLP confirmed that the chickpea isolates should be divided into 2 major phylogenetic groups including *Mesorhizobium ciceri* (formerly *Rhizobium ciceri*) and a novel species *Mesorhizobium mediterraneum* (formerly *Rhizobium mediterraneum*) [25]. The 16S–IGS RFLP could differentiate the strains nodulating *V. faba* from defined *Rhizobium* species with similarities less than 72.8% and supported the proposal of a novel species, *Rhizobium fabae* [26].

6.6. PCR-RFLP OF THE SYMBIOTIC GENES OF RHIZOBIA

The symbiotic genes have been used as markers of rhizobial populations [17, 18]. Primers FGP807 and FGP4929 were used to amplify IGS regions between *nifD* and *nifH* genes and parts of these genes, which are located on symbiotic plasmids in most *Rhizobium* strains. The primer pair NODD2PH678 and NODD3PH2152' amplified *nodD2* through *nodD3* fragments which include *nodD2*, the IGS sequence between *nodD2* and *nodD3*, and about 120 bp of *nodD3* in *R. leguminosarum* biovar phaseoli. All *R. leguminosarum* biovar phaseoli strains produced a single band of about 1,490 bp [17]. The primer pair NBA12 and NBF12 amplified *nodD* through *nodF* fragments including *nod* and the IGS sequence between *nodD* and *nodF* in *R. leguminosarum* biovars viciae and trifolii. *R. leguminosarum* biovar trifolii strains produced a single band of about 1,450 bp. The size of the *nodD-F* PCR products of *R. leguminosarum* biovar viceae varied from approximately 1,350 to 2,200 bp [18]. In RFLP of 5 parts of the symbiotic region adjacent to *nod* box sequences, the *nod* box regions were highly conserved among strains belonging to each of the 2 biovars but differ structurally to various degrees between the biovars [27].

PCR-RFLP analysis of symbiotic gene regions grouped the strains within *R. leguminosarum* species and within biovars independently of their chromosomal types. Intrabiovar variation within symbiotic gene regions was detected by PCR-RFLP analysis of *nifDK* and *nodD* regions [17]. The PCR-RFLP analysis of *nifH* and *nodC* genes resulted in a classification of rhizobia which was in general well-correlated with their known host range and independent of their taxonomic status [28].

6.7. ADVANTAGES AND LIMITATIONS OF PCR-RFLP

PCR-RFLP has become valuable because of the following characteristics: i) this method is rapid, simple, sensitive, reliable, reproducible, easily analyzable and suitable for estimating the genetic relationships among rhizobial populations at species higher and below levels; ii) it can be used for a rapid tracking of the known relatives of new isolates; iii) several specific regions can be chosen to provide more alternatives which are suitable for any particular population; iv) combination of the banding pattern across different enzyme digests can be applied for the identification of rhizobia and the detection of new taxa; v) a dendrogram can be constructed from the combined patterns to estimate the genetic relatedness among rhizobial populations; vi) it is useful in rhizobial taxonomy since the results obtained from the 16S PCR-RFLP are in good agreement with those from 16S rDNA sequence analysis, DNA-DNA hybridization, MLEE and numerical taxonomy; vii) the results obtained from IGS PCR-RFLP are consistent with those obtained from amplified fragment length polymorphism (AFLP) and ARDRA, but have higher discriminating power; viii) the results obtained from IGS PCR-RFLP correspond reasonably well to the rhizobial classification; ix) intrabiovar variation of rhizobia can be detected by PCR-RFLP analysis of symbiotic region; x) in some case, less number of endonucleases can provide the maximum discriminating power. The limitations of PCR-RFLP include i) it is less discriminating than DNA-DNA hybridization and MLEE; ii) PCR-RFLP of symbiotic regions is less valuable for taxonomic purpose; iii) in some case, the sequence divergence is not particularly high; iv) a mutation at restriction site may cause different patterns.

Table 1: Genes and endonucleases frequently used in PCR-RFLP

Amplified Gene	Endonuclease Used	References
16S rDNA	*Alu*I, *Cfo*I, *Dde*I, *Hae*III, *Hin*fI, *Msp*I, *Nde*II, *Rsa*I and *Taq*I	[1]
	*Alu*I, *Dde*I, *Hae*III, *Hha*I, *Hin*fI, *Msp*I, *Nde*II, *Rsa*I and *Taq*I	[16]
	*Cfo*I, *Hin*fI, *Msp*I, *Nde*II and *Rsa*I	[5]
	*Alu*I, *Hae*III, *Mbo*I and *Msp*I	[19]
	*Alu*I, *Hae*III, *Msp*I and *Nde*II	[29]
	*Hae*I, *Hin*fI, *Msp*I and *Rsa*I	[14]
	*Hha*I, *Hin*fI, *Msp*I and *Sau*3AI	[8-10, 13, 30]
	*Cfo*I, *Hin*fI and *Msp*I	[15]
	*Msp*I, *Pst*I and *Sau*3AI	[31]
	*Hae*III and *Hin*fI	[7]
23S rDNA	*Alu*I, *Cfo*I, *Dde*I, *Hae*III, *Hin*fI, *Mbo*I, *Mse*I, *Msp*I and *Rsa*I	[3]
	*Alu*I, *Hae*III, *Mbo*I and *Msp*I	[19]
	*Cfo*I, *Hin*fI and *Msp*I	[15]
16S-23S IGS	*Alu*I, *Cfo*I, *Dde*I, *Hae*III, *Hin*fI, *Mbo*I, *Mse*I, *Msp*I and *Rsa*I	[3]
	*Alu*I, *Cfo*I, *Dde*I, *Hae*III, *Hin*fI, *Msp*I, *Nde*II and *Taq*I	[17]
	*Alu*I, *Dde*I, *Hae*III, *Hha*I, *Hin*fI, *Msp*I, *Nde*II and *Rsa*I	[21]

Table 1: cont….

	*Alu*I, *Cfo*I, *Dde*I, *Hae*III, *Msp*I and *Nde*II	[18]
	*Cfo*I, *Hae*III and *Msp*I	[22]
16S rDNA and 16S-23S IGS	*Alu*I, *Cfo*I, *Hae*III, *Msp*I, *Nde*I, *Rsa*I and *Sau*96.1	[25]
	*Hin*fI, *Msp*I and *Rsa*I	[23]
nifH	*Hae*III, *Msp*I and *Nde*I	[28]
nifD through *nifK*	*Alu*I, *Cfo*I, *Dde*I, *Hae*III, *Hin*fI, *Msp*I, *Nde*II and *Taq*I	[17]
nodC	*Cfo*I, *Hae*III, *Hin*fI, *Msp*I, *Nde*II and *Rsa*I	[28]
nodD through *nodF*	*Alu*I, *Cfo*I, *Dde*I, *Hae*III, *Hin*fI, *Msp*I, *Nde*II and *Taq*I	[17]
	*Alu*I, *Cfo*I, *Dde*I, *Hae*III, *Msp*I and *Nde*II	[18]
nodD2 through *nodD3*	*Alu*I, *Cfo*I, *Dde*I, *Hae*III, *Hin*fI, *Msp*I, *Nde*II and *Taq*I	[17]
	*Alu*I, *Cfo*I, *Dde*I, *Hae*III, *Msp*I and *Nde*II	[18]
nod box region	*Alu*I, *Hae*III and *Msp*I	[27]

16S rDNA: small subunit ribosomal RNA gene; 23S rDNA: large subunit ribosomal RNA gene; 16S-23S IGS: intergenic spacer between 16S and 23S rDNA; *nif*: nitrogen fixation gene; *nod*: nodulation gene

Table 2: Length of 16S-23S IGS region in rhizobia

Rhizobial Species	Length of IGS Region (bp)	References
Azorhizobium spp.	1,000	[24]
Bradyrhizobium elkanii	852 to 870	[21]
Bradyrhizobium japonicum	300; 700; 800 to 900	[21, 23, 24]
Bradyrhizobium liaoningense	780; 800	[23]
Bradyrhizobium spp.	780 to 1,038	[21, 23, 24]
Ensifer fredii (formerly *Rhizobium fredii*)	1,300	[24]
Ensifer medicae (formerly *Sinorhizobium medicae*)	1,200	[23]
Ensifer meliloti (formerly *Sinorhizobium meliloti*)	1,200; 1,300	[23, 24]
Ensifer terangae (formerly *Sinorhizobium terangae*)	1,200	[23]
Mesorhizobium ciceri (formerly *Rhizobium ciceri*)	900; 1,000	[23, 24]
Mesorhizobium huakuii	800	[23]
Mesorhizobium loti (formerly *Rhizobium loti*)	1,000	[24]

Table 2: cont....

Mesorhizobium mediterraneum (formerly *Rhizobium mediterraneum*)	1,000	[25]
Mesorhizobium tianshanense	800	[23]
Rhizobium etli	1,000	[23]
Rhizobium galegae	1,000	[24]
Rhizobium hainanense	1,200	[23]
Rhizobium huautlense	1,100	[23]
Rhizobium leguminosarum	1,160 to 1,400	[17, 23, 24]
Rhizobium mongolense	1,200	[23]
Rhizobium tropici	1,000; 1,300	[23]
Rhizobium undicola (formerly *Allorhizobium undicola*)	1,300	[23]
Rhizobium spp.	1,000	[25]

bp: base pairs.

REFERENCES

[1]   Laguerre G, Allard M, Revoy F, Amarger N. Rapid identification of rhizobia by restriction fragment length polymorphism analysis of PCR-amplified 16S rRNA genes. Appl Envir Microbiol 1994; 60: 56-63.

[2]   Ralph D, McClelland M, Welsh J, *et al. Leptospira* species categorized by arbitrarily primed polymerase chain reaction (PCR) and by mapped restriction polymorphisms in PCR-amplified rRNA genes. J Bacteriol 1993; 175: 973-81.

[3]   Terefework Z, Nick G, Suomalainen S, *et al.* Phylogeny of *Rhizobium galegae* with respect to other rhizobia and agrobacteria. Int J Syst Bacteriol 1998; 48: 349-56.

[4]   Weisburg WG, Barns SM, Pelletier DA, Lane DJ. 16S ribosomal DNA amplification for phylogenetic study. J Bacteriol 1991; 173: 697-03.

[5]   Grange L, Hungria M. Genetic diversity of indigenous common bean (*Phaseolus vulgaris*) rhizobia in two Brazilian ecosystems. Soil Biol Biochem 2004; 36: 1389-98.

[6]   Nei M, Li WH. Mathematical model for studying genetic variation in terms of restriction endonucleases. Proc Natl Acad Sci USA. 1979; 76: 5267-73.

[7]   Elboutahiri N, Thami-Alami I, Zaid E, Udupa SM. Genotypic characterization of indigenous *Sinorhizobium meliloti* and *Rhizobium sullae* by rep-PCR, RAPD and ARDRA analyses. Afr J Biotechnol 2009; 8: 979-85.

[8]   Wang ET, van Berkum P, Sui XH, *et al.* Diversity of rhizobia associated with *Amorpha fruticosa* isolated from Chinese soils and description of *Mesorhizobium amorphae* sp. nov. Int J Syst Bacteriol 1999; 49: 51-65.

[9]   Diouf A, de Lajudie P, Neyra M, *et al.* Polyphasic characterization of rhizobia that nodulate *Phaseolus vulgaris* in West Africa (Senegal and Gambia). Int J Syst Evol Microbiol 2000; 50: 159-70.

[10]   Wang ET, Rogel MA, Garcia-de los Santos A, *et al. Rhizobium etli* bv. mimosae, a novel biovar isolated from *Mimosa affinis*. Int J Syst Bacteriol 1999; 49: 1479-91.

[11] Laguerre G, Fernandez MP, Edel V, *et al.* Genomic heterogeneity among French *Rhizobium* strains isolated from *Phaseolus vulgaris* L. Int J Syst Bacteriol 1993; 43: 761-67.

[12] Tan Z, Wang ET, Peng GX, *et al.* Characterization of bacteria isolated from wild legumes in the north-western regions of China. Int J Syst Bacteriol 1999; 49: 1457-69.

[13] Wang ET, van Berkum P, Beyene D, *et al. Rhizobium huautlense* sp. nov., a symbiont of *Sesbania herbacea* that has a close phylogenetic relationship with *Rhizobium galegae*. Int J Syst Bacteriol 1998; 48: 687-99.

[14] Wei GH, Wang ET, Tan ZY, *et al. Rhizobium indigoferae* sp. nov. and *Sinorhizobium kummerowiae* sp. nov., respectively isolated from *Indigofera* spp. and *Kummerowia stipulacea*. Int J Syst Evol Microbiol 2002; 52: 2231-39.

[15] Shamseldin A, El-Saadani M, Sadowsky MJ, An CS. Rapid identification and discrimination among Egyptian genotypes of *Rhizobium leguminosarum* bv. *viciae* and *Sinorhizobium meliloti* nodulating faba bean (*Vicia faba* L.) by analysis of *nodC*, ARDRA, and rDNA sequence analysis. Soil Biol Biochem 2009; 41: 45-53.

[16] Lafay B, Burdon JJ. Molecular diversity of rhizobia occurring on native shrubby legumes in southeastern Australia. Appl Envir Microbiol 1998; 64: 3989-97.

[17] Laguerre G, Mavingui P, Allard MR, *et al.* Typing of rhizobia by PCR DNA fingerprinting and PCR-restriction fragment length polymorphism analysis of chromosomal and symbiotic gene regions: application to *Rhizobium leguminosarum* and its different biovars. Appl Envir Microbiol 1996; 62: 2029-36.

[18] Laguerre G, Louvrier P, Allard M, Amarger N. Compatibility of rhizobial genotypes within natural populations of *Rhizobium leguminosarum* biovar viciae for nodulation of host legumes. Appl Envir Microbiol 2003; 69: 2276-83.

[19] Wolde-meskel E, Terefework Z, Frostegard A, Lindstrom K. Genetic diversity and phylogeny of rhizobia isolated from agroforestry legume species in southern Ethiopia. Int J Syst Evol Microbiol 2005; 55: 1439-52.

[20] Thorsson AT, Sverrisson H, Anamthawat-Jonsson K. Genotyping Icelandic isolates of rhizobia based on rDNA-RFLP. Buvisindi Icel Agr Sci 2000; 13: 17-25.

[21] Doignon-Bourcier F, Willems A, Coopman R, *et al.* Genotypic characterization of *Bradyrhizobium* strains nodulating small Senegalese legumes by 16S-23S rRNA intergenic gene spacers and amplified fragment length polymorphism fingerprint analyses. Appl Envir Microbiol 2000; 66: 3987-97.

[22] Gu CT, Wang ET, Tian CF, *et al. Rhizobium miluonense* sp. nov., a symbiotic bacterium isolated from *Lespedeza* root nodules. Int J Syst Evol Microbiol 2008; 58: 1364-68.

[23] Quatrini P, Scaglione G, Cardinale M, *et al. Bradyrhizobium* sp. nodulating the Mediterranean shrub Spanish broom (*Spartium junceum* L.). J Appl Microbiol 2002; 92: 13-21.

[24] Nour SM, Fernandez MP, Normand P, Cleyet-Marel JC. *Rhizobium ciceri* sp. nov., consisting of strains that nodulate chickpeas (*Cicer arietinum* L.). Int J Syst Bacteriol 1994; 44: 511-22.

[25] Nour SM, Cleyet-Marel JC, Normand P, Fernandez MP. Genomic heterogeneity of strains nodulating chickpeas (*Cicer arietinum* L.) and description of *Rhizobium mediterraneurn* sp. nov. Int J Syst Bacteriol 1995; 45: 640-48.

[26] Tian CF, Wang ET, Wu LJ, *et al. Rhizobium fabae* sp. nov., a bacterium that nodulates *Vicia faba*. Int J Syst Evol Microbiol 2008; 58: 2871-75.

[27]   Andronov EE, Terefework Z, Roumiantseva ML, *et al.* Symbiotic and genetic diversity of *Rhizobium galegae* isolates collected from the *Galega orientalis* gene center in the Caucasus. Appl Envir Microbiol 2003; 69: 1067-74.

[28]   Laguerre G, Nour SM, Macheret V, *et al.* Classification of rhizobia based on *nodC* and *nifH* gene analysis reveals a close phylogenetic relationship among *Phaseolus vulgaris* symbionts. Microbiol 2001; 147: 981-93.

[29]   Aguilar OM, Lopez MV, Riccillo PM, *et al.* Prevalence of the *Rhizobium etli*-like allele in genes coding for 16S rRNA among the indigenous rhizobial populations found associated with wild beans from the Southern Andes in Argentina. Appl Envir Microbiol 1998; 64: 3520-24.

[30]   Yao ZY, Kan FL, Wang ET, *et al.* Characterization of rhizobia that nodulate legume species of the genus *Lespedeza* and description of *Bradyrhizobium yuanmingense* sp. nov. Int J Syst Evol Microbiol 2002; 52: 2219-30.

[31]   Silva C, Vinuesa P, Eguiarte LE, *et al. Rhizobium etli* and *Rhizobium gallicum* nodulate common bean (*Phaseolus vulgaris*) in a traditionally managed Milpa Plot in Mexico: population genetics and biogeographic implications. Appl Envir Microbiol 2003; 69: 884-93.

CHAPTER 7

Genotypic Diversity of Rhizobia Assessed by Restriction Fragment Length Polymorphism (RFLP)

Neelawan Pongsilp[*]

Department of Microbiology, Faculty of Science, Silpakorn University, Nakhon Pathom, Thailand

Abstract: Restriction fragment length polymorphism (RFLP) has been used to examine the genotypic diversity of bacteria. The technique is based on restriction analysis and hybridization, resulting in the fingerprint patterns. For rhizobia, symbiotic genes have been frequently used as probes for hybridization. Based on the same approach, the insertion sequences (ISs) can be used as probes, resulting in the patterns termed as "IS fingerprints". The use of ISs as probes for hybridization can provide high-resolution fingerprints of rhizobial strains. The ISs have been found to be abundant in rhizobia. The distribution of ISs varies widely in both IS type and copy number. Several ISs are specific to rhizobial species. The potential of ISs to discriminate among different rhizobial isolates has been reported.

Keywords: Genotypic diversity, Hybridization, Insertion sequence (IS), IS fingerprint, Repeated sequence (RS), Restriction fragment length polymorphism (RFLP), Rhizobia, Southern blot, Symbiotic gene, Transposition.

7. 1. PRINCIPLE OF RESTRICTION FRAGMENT LENGTH POLYMOR-PHISM (RFLP)

Restriction fragment length polymorphism (RFLP) approaches in determining the genetic relationships are based on restriction analysis and hybridization. In the protocol, total genomic DNA is digested with endonuclease (restriction enzyme). The digested DNA fragments are separated in 1.0% agarose gel electrophoresis, then are transferred to a nitrocellulose membrane or nylon membrane by the Southern blot procedure. The restriction patterns of total genomic DNA are very complex, thus it is very difficult to distinguish among strains unless the differences in the restriction patterns are marked. Hybridization of the membrane

*Address correspondence to Neelawan Pongsilp:** Department of Microbiology, Faculty of Science, Silpakorn University, Nakhon Pathom, Thailand; Tel: +66-34-245337; Fax: +66-34-245336-37; E-mail: neelawan@su.ac.th

with a labeled DNA probe enables the determination of the fragments which are complementary to the probe. Southern blot and hybridization can be carried out as described previously [1, 2]. To prepare a labeled DNA probe, the gene fragment, obtained from either restriction enzymatic digestion or PCR amplification, is labeled with either radioactives (such as ^{125}I and ^{32}P) or chemicals (such as digoxigenin and biotin) by ramdom oligonucleotide primed synthesis. The differences in genomes can be detected by the presence of fragments in different lengths, in which a gene complementary to probe resides, resulting in the fingerprint patterns termed as "RFLP".

7.2. RFLP FOR EXAMINING GENOTYPIC DIVERSITY OF RHIZOBIA

RFLP with probes of chromosomal DNA fragments and symbiotic genes provide useful genetic markers to examine the genetic diversity of rhizobia. The presence of symbiotic genes has been detected by DNA-DNA hybridization. Symbiotic gene probes including *nif* (nitrogen fixation) and common *nod* (nodulation) genes have been used previously to characterize *Rhizobium* and *Bradyrhizobium* [3-9]. The *nif* genes are required for nitrogenase activity [10, 11]. The *nodABC* genes are involved in the synthesis of N-acytylated chitooligosaccharides, the backbone core structures of Nod factor molecules. The *nodD* gene encodes a transcriptional regulator of nodulation genes. The *nodABCD* genes have been found in all rhizobia studied so far and have been referred to as common *nod* genes [12, 13]. The common *nod* gene region is a cluster of *nod* genes that are structurally and functionally conserved in all rhizobial strains. DNA sequencing has revealed a high degree of homology (about 70%) in the amino acid sequences of corresponding common *nod* genes from different *Rhizobium* strains [10]. Sadowsky *et al.* [7] reported that there was a relationship between RFLP with *nif* probe of 20 *Bradyrhizobium japonicum* isolates and nodulation classes since *nif* RFLP patterns fell into 3 markedly different groups, which correlated well with the nodulation classes. While there was no relationship between RFLP with *nod* probe and nodulation classes or the groupings based on *nif* gene hybridization. RFLP with probes of chromosomal DNA fragments and symbiotic genes (*nif* and *nod* genes) was used to examine the genetic diversity of 18 strains of *Rhizobium leguminosarum* biovar trifolii, a strain of *R. leguminosarum* biovar viciae and 2 strains of *Ensifer meliloti* (formerly *Rhizobium meliloti*). Analysis of RFLP around chromosomal DNA probes also delineated 16 unique RFLP patterns and yielded a genetic diversity similar to that revealed by the allozyme data. Analysis of RFLP around symbiotic plasmid-derived probes demonstrated that the symbiotic plasmids reflect genetic divergence similar to that of their bacterial hosts. These markers can facilitate studies of the ecology of the symbiotic plasmids in nature [13]. Hybridization of total genomic DNA from *R. leguminosarum* with a probe of *cpn60* gene, encoding chaperonin from *Escherichia coli*, also showed evidence for at least 2 homologs of *cpn60* [14]. Paffetti *et al.* [15] investigated the genetic diversity of *Ens. meliloti* populations and found that RFLP analysis of symbiotic *nod/nol* operon was consistent with the result obtained from random amplified polymorphic DNA (RAPD). The isolates that nodulate *Phaseolus vulgaris* (common bean) were found to contain 3 copies of the *nifH* gene since 3 restriction fragments ranging in sizes from 3.5 to 9.0 kilobases (kb) hybridized to the *nifH* probe [16]. RFLP with *nodAB, nodCIJT* and the 1.0-kb reiterated chromosomal sequence was used to estimate the level of symbiotic plasmid transfer among different chromosomal lineages of *Rhizobium* nodulating *Trifolium* (clover). A strict association existing between major symbiotic plasmid and chromosomal genetic group was observed, indicating a lack of successful transfer of nodulation genes between different chromosomal groups [17]. DNA sequences homologous to the genes *fixLJKNOPQGHIS* not present in the symbiotic plasmid were found on the chromosome of *Ensifer fredii* (formerly *Rhizobium* sp.) NGR234, while sequences homologous to *nodPQ* and *exoBDFLK* were found on its magaplasmid [18]. Hybridization to *nif, nod* and *fix* gene probes indicated that most of the symbiotic genes could be transferred from the symbiotic plasmid to the nonsymbiotic plasmid in *R. leguminosarum* biovar viciae. On the other hand, some DNA fragments originating from the nonsymbiotic plasmid transferred to the symbiotic plasmid. These results suggest the occurrence of a reciprocal but unequal DNA exchange between the 2 plasmids [19]. The *nifH* hybridization patterns also revealed the occurrence of lateral plasmid transfer within species of *Rhizobium etli* and *Rhizobium gallicum* but not between species [20]. DNA-DNA hybridization was used to determine the relationships between *Lespedeza* rhizobia and the related species [21]. RFLP patterns with the insertion sequence (IS) probes (IS*Rm1*, IS*Rm3* and IS*Rm5*) and the symbiotic gene probes (*nifH, nodC, nodEFG* and *nodH*) were accomplished to identify rhizobia from nodules of *Medicago sativa* (alfalfa) and *Melilotus alba* (sweet clover). RFLP with *nodH* and *nodEFG* probes divided the tested strains into 6 and 9 *nod* genotypes, respectively, whereas RFLP with the 3 IS probes divided the strains into 8 IS genotypes [22]. DNA-DNA hybridization was also developed for quality control of *Rhizobium* inoculants. The hybridization probes were total DNA from pure cultures of the inoculants strains *Rhizobium galegae* and *R. leguminosarum* as well as a 264-base pairs (bp) strain-specific fragment from the genome of *R. galegae*. The total DNA probes distinguished inoculants containing *R. galegae* or *R. leguminosarum*, and the strain-specific probe distinguished inoculants containing *R. galegae*. Therefore, when suitable probes are available, the method offers a promising alternative for the quality control of peat-based inoculants [23]. DNA probes specific for *Mesorhizobium loti* (formerly *Rhizobium loti*), *R. leguminosarum* biovar phaseoli and *Rhizobium tropici* groups have been developed [24, 25]. DNA-DNA hybridization offers great benefits for the development of DNA probes for monitoring bacterial populations in environmental samples [25].

7.3. INSERTION SEQUENCE (IS) FINGERPRINTING OF RHIZOBIA

7.3.1. Distribution of ISs in Rhizobia

Inserion sequences (ISs) are small genetic elements able to transpose from a donor replicon to a new location on the genome [26]. ISs are constituted basically by a gene that encodes the transposase (an enzyme required for transposition) and inverted repeat sequences (IRs) at both sides. ISs are present in the genomes, plasmids and bacteriophages of a wide range of bacterial genera and species. Copies of ISs can be located on both chromosome and plasmids of the same bacterial cell. Some ISs probably move from plasmids to chromosome, but they do not appear to move to other plasmids. ISs have been grouped in different families according to their similarity, transposase type, transposition mechanism and the sequence of IRs. The representation of ISs families varies widely among bacteria. Some species lack ISs, whereas others contain more than 100 ISs of different families or belong to only one predominant family. Neither the origin of ISs nor their cellular function is known, except for some of them that code for adaptive phenotypes such as antibiotic resistance. In rhizobia, ISs are fairly abundant in the symbiotic plasmids or chromosomal islands that carry genes essential for symbiosis. The ISs from the symbiotic plasmids were frequently present in the analyzed strains, whereas the chromosomal ISs were observed less frequently. The ISs presented within the symbiotic plasmid have a higher degree of genomic context conservation and lower nucleotide diversity [27]. The numerous ISs in the genomes of rhizobia were already sequenced. ISs compose another subclass of repetitive sequences from rhizobia [28].

ISs have been investigated for their distribution in rhizobial strains by hybridization using IS probes. It is generally believed that ISs in rhizobia promote genetic diversification through genomic rearrangement and recombination [29-31]. The use of ISs as probes for hybridization can provide high-resolution fingerprints of rhizobial strains [32, 33]. The IS*Rm1* is normally present in about 10 copies per genome of *Ens. meliloti* 1021 and this IS was further characterized by DNA sequence analysis [34]. The presence of IS*Rm1* is considered as quite common characteristics for *Ens. meliloti* because approximately 80% of *Ens. meliloti* strains contain 1 to at least 11 copies of IS*Rm1* in their genomes [35]. The IS*Rm2* in *Ens. meliloti* was found to carry the symbiotic *fixX* gene encoding a ferredoxin, a protein that mediates electron transfer and provides energy for nitrogen fixation [36]. The IS*Rm3* and IS*Rm4* were discovered in different *Ens. meliloti* strains. The IS*Rm3* homologs were subsequently detected in *R. leguminosarum* and over 90% of *Ens. melioti* strains [37, 38]. The IS*Rm4* homologs were found in *R. leguminosarum* and *Ens. fredii* (formerly *Sinorhizobium fredii*). The frequency of IS*Rm3* and IS*Rm4* homologs within each *R. leguminosarum* biovar viciae population tested was also variable. The IS*Rm4* is closely related to IS*Rl2* which was recently found in *R. leguminosarum* biovar viciae [37]. A target for IS*Rm3* transposition in *Ens. meliloti* IZ450 is another IS, named IS*Rm5*. Multiple copies and variants of IS*Rm5* that occur in *Ens. meliloti* genome are often in a close association with IS*Rm3* [30]. More than 25 distinct ISs have been found in *Ens. meliloti*, and it is common for several ISs to occupy the same genome in multiple copies [26]. The 16 ISs from *Ens. meliloti* were further used to characterize various *Ens. meliloti* strains by hybridization. The resulting hybridization patterns were different for every strain and gave a clear and definite IS fingerprint of each strain. These IS fingerprints can be used to identify and characterize *Ens. melioti* strains rapidly and unequivocally, as they proved to be relatively stable. The IS element which gave the most detailed fingerprint was IS*Rm2011-2* [33]. The IS*Rm2011-2* was widely distributed and abundant within *Ens. meliloti* indigenous populations, being found in all tested strains with copy number varying from 1 to 18 [39, 40]. The IS homologous to IS*Rl2* was found within strains belonging to *Ens. meliloti* (formerly *Sinorhizobium meliloti*), *Ens. fredii*, *R. etli*, *R. galegae*, *R. leguminosarum*, *R. tropici* and *Rhizobium* sp. The apparent copy number of IS*Rl2* varies from 1 to 8 [26]. The 39 ISs are present in the genome of *R. etli* CFN42 [41]. The IS*Rel4* and IS*Rel2* (from the symbiotic plasmid of *R. etli* CFN42) were the most common ISs, present in more than 50% of *R. etli* popualtions [27]. The 2 different repeated sequences (RSs), RSRjα and RSRjβ, were discovered in *B. japonicum* (formerly *Rhizobium japonicum*) genome. Many copies of both RSs are clustered around the nitrogenase genes *nifDK* and *nipfl*. RSα and RSβ possess structural characteristics of ISs [42]. The 21 isolates of *B. japonicum* showing numerous RS-specific hybridization bands were designated highly reiterated sequence-possessing (HRS) isolates. Some HRS isolates possessed extremely high numbers of RSα copies, ranging from 86 to 175 [43]. The distribution of ISs and IS homologs in rhizobial species is shown in Table **1**.

7.3.2. IS Fingerprint for Examining Genotypic Diversity of Rhizobia

IS fingerprints have been used for examination of genotypic diversity of rhizobia. The 61 isolates of *Ens. meliloti* were placed into 24 IS classes based on IS*Rm3* and IS*Rm5* hybridization [44]. The diversity of *Ens. meliloti* nodulating *Med. sativa* isolated from acid soils of Argentina and Uruguay was assayed by IS fingerprinting techniques using an IS*Rm2011-2* probe. The dendrogram constructed from IS fingerprints revealed a high diversity among the isolates but some of them appeared related to inoculant strains used in the region [40]. The IS probe IS*RLd*TAL1145-1 from *Rhizobium* sp. TAL 1145 hybridized strongly to genomic DNA from 10 rhizobial strains that nodulate both *Leucaena leucocephala* and *Pha. vulgaris*. A collection of strains, including a strain of *Rhizobium* sp. that nodulates *Leucaena* spp., 9 strains of *Ens. meliloti* that nodulate *Med. sativa*, 4 strains of *Rhizobium* spp. that nodulate *Sophora chrysophylla*, and a non-nodulating bacterium associated with the nodules of *Pithecellobium dulce* from the *Leucaena* cross-inoculation group, produced distinguishing IS patterns for each strain [45]. The IS*Rm1*, IS*Rm3* and IS*Rm5* probes were used for hybridization of rhizobia from nodules of *Med. sativa* and *Mel. alba* which including rhizobia in the genera *Ensifer*, *Phyllobacterium* and *Rhizobium*. The profiles obtained from the 3 IS probes divided the strains into 8 IS genotypes [22]. The potential of IS*Rl2* to discriminate among different isolates of *Ens. fredii* has been reported [26].

The consistency has been reported for IS fingerprints and the other molecular approaches. Grouping of 21 isolates of *Ens. meliloti* based on their IS fingerprints correlated with grouping of strains based on their RAPD and enterobacterial repetitive intergenic consensus-polymerase chain reaction (ERIC-PCR) fingerprints [46]. *R. galegae* strains were clearly distinguished and grouped within each host plant based on IS fingerprints, PCR-RFLP of internal transcribed spacer of rRNA gene (ITS) and PCR-RFLP of *nod* box. It has been noticed that all strains containing ISs had closely similar ITS types. Strains with identical IS fingerprints had identical amplified fragment length polymorphism (AFLP) fingerprints. The IS has been found only in strains which have genetically similar backgrounds, indicating a limitation of horizontal transfer between different chromosomal groups [47]. On the other hand, horizontal gene transfer was found with IS*Rm1* and IS*Rm2* from *Ens. meliloti* to *Agrobacterium tumefaciens* [48]. The IS homologs are present in rhizobia which have distantly related genomes, suggesting that *Rhizobium* IS elements are prone to genetic transfer [49].

7.3.3. IS Fingerprint for Identification of Rhizobial Species

Hybridization with IS probes has been developed for a positive identification of rhizobial strains. The plasmid pRWRm13 containing 900-bp internal *Bgl*I-*Eco*RV fragment of IS*Rm1* in plasmid vector, pUC9, was constructed. The patterns produced by hybridization of the probe to *Ens. meliloti* genomic DNA digested with a series of endonucleases were reproducible and distinctive for each strain. Therefore, IS*Rm1* can be used for *Ens. meliloti* strain identification. Strain identification by probing with IS*Rm1* is applicable to *Ens. meliloti* containing IS*Rm1* from any source including field isolates, nodules or strain collections. It may also be used as a stringent adjunct to more rapid strain identification methods for studies requiring the screening of a large number of isolates [50].

7.3.4. Relation Between ISs and Symbiotic Genes

ISs are fairly abundant in plasmids or chromosomal islands that carry the genes needed for symbiosis. ISs contained within the symbiotic plasmid have a higher degree of genomic context conservation, lower nucleotide diversity and genetic differentiation, and fewer recombination events than the chromosomal housekeeping genes [27]. ISs found in rhizobia are of particular interest because of their potential for interaction with genes required for symbiotic nitrogen fixation [35]. A close association was found in IS*Rm2* of *Ens. meliloti* that carries the symbiotic gene *fixX* [36]. It has been hypothesized that ISs may represent common control elements for symbiotic genes. The IS*Rm1* has been shown to transpose at high frequency into *fix* genes on the *nod* megaplasmid of *Ens. meliloti* 1021 [34] and has been found to be important as a widespread and frequently-occurring mutagen capable of disrupting genes which are essential for the symbiosis [35]. IS*Rl2* of *R. leguminosarum* has been found to be associated with the symbiotic plasmid [26]. RSRjα and RSRjβ of *B. japonicum* are clustered around *nifDK* and *nifH* gene, but not involved in the expression of *nifDK* [42]. Transposition of IS*Rm2* into nitrogen fixation and nodulation genes located on the symbiotic plasmid of *Ens. meliloti* was detected at a high frequency. However, the arrangement of IS*Rm2* copies was identical in the free-living cells and in nitrogen-fixing nodules, suggesting that the involvement of IS*Rm2* transposition in the development of nitrogen-fixing symbiosis is unlikely [36].

7.4. ADVANTAGES AND LIMITATIONS OF RFLP

RFLP offers great benefits because of the following characteristics: i) this method provides stable and reproducible results; ii) it provides useful genetic markers to examine the genetic diversity and to characterize the symbiotic genes of rhizobial strains; iii) it provides a promising alternative for tracking rhizobial species and for monitoring bacterial populations in environmental samples; iv) RFLP with suitable probes can discriminate among different isolates within rhizobial species; v) IS fingerprints can provide high-resolution fingerprints of rhizobial strains; vi) IS fingerprints with species-specific probes can be used for identification of rhizobial species; vii) RFLP with symbiotic probes can facilitate studies of the ecology of the symbiotic plasmids in nature; viii) a dendrogram can be constructed from RFLP patterns to present the genetic relatedness among rhizobial populations; ix) RFLP with chromosomal DNA probe has been found to be consistent with the result obtained from the allozyme data; x) RFLP with symbiotic probes has been found to be consistent with the result obtained from RAPD; xi) IS fingerprinting has been found to be consistent with the results obtained from other molecular techniques including RAPD, ERIC-PCR, PCR-RFLP of ITS and AFLP. The limitations of RFLP are that i) it consists of several procedures that are time- and labor-consuming, requiring fastidious DNA extraction, probe preparation, hybridization and detection; ii) a mutation at restriction site may cause different patterns.

Table 1: Distribution of insertion sequences (ISs) and IS homologs in rhizobial species

Rhizobial Species	IS Name	IS Family	Length (bp)	References
Bradyrhizobium japonicum (formerly *Rhizobium japonicum*)	HRS1	IS*1380*	2,071	[51]
	IS*B20*	-	1,809	[52]
	IS*1631*	IS3 (IS*21*)	2,712	[52]
	IS*1632*	IS256	1,395	[52]
	RSα	IS630-Tc1	1,122	[42, 43, 53]
	RSβ	IS3	950	[42, 53]
	RSγ	IS*NCY*	1,000	[53]
	RSδ	IS*NCY*	1,000	[53]
	RSε	IS*NCY*	1,000	[53]
	RSζ	IS*NCY*	2,000	[53]
Bradyrhizobium sp.	IS*R1*	IS3	1,150	[54]
	unnamed IS	IS3	998	[55]
Ensifer fredii (formerly *Sinorhizobium fredii* and *Rhizobium fredii*) (*Ens. fredii* NGR234 was formerly *Rhizobium* sp. NGR234)	IS*NGR3*	IS*21*	2,625	[49]
	IS*NGR4*	-	3,324	[49]
	IS*Rel4*	IS*As1*	-	[27]
	IS*Rel9*	IS5	1,571	[27]
	IS*Rf1*	Tn*3*	1,002	[56]
	IS*R12* homolog	IS5	932	[26]
	IS*Rm4* homolog	IS5	nd	[37]
	IS*Rm102F34-1*	-	1,481	[32]
	IS*Rsp1*	IS66	3,481	[57]
Ensifer meliloti (formerly *Sinorhizobium meliloti* and *Rhizobium meliloti*)	IS*RLd* TAL1145-1	IS*NCY*	2,500	[45]
	IS*R12* homolog	IS5	932	[26]
	IS*Rm*MVII-1	-	4,200	[33]

Table 1: cont...

	ISRmMVII-2	-	1,800	[33]
	ISRmMVII-3 (ISRm USDA1024-1)	-	2,700	[33]
	ISRmMVII-4	-	3,800	[33]
	ISRmMVII-10	-	5,750	[33]
	ISRm USDA1024-2 (ISRm2011-1)	-	1,400	[33]
	ISRm USDA1024-3	-	3,000	[33]
	ISRm1	IS3	1,319	[34, 35, 49]
	ISRm2	IS66	2,700	[36]
	ISRm3	IS256	1,298; 1,316	[37, 38, 44, 58]
	ISRm4	IS5	933	[37, 59]
	ISRm4-1	IS5	936	[60]
	ISRm5	IS256	1,340	[30, 44, 61, 62]
	ISRm6	IS3	1,269	[37, 63]
	ISRm8	IS701	1,451	[64]
	ISRm9	IS21	2,797	[60]
	ISRm9 derivative	IS21	4,787	[65]
	ISRm10	IS630-Tc1/IS3	1,047	[66]
	ISRm10-1	IS630-Tc1	1,042	[67]
	ISRm10-2	IS630-Tc1	1,049	[67]
	ISRm12	-	>1,484	[65]
	ISRm14	IS66	2,695	[68]
	ISRm15 derivative	-	5,398	[65]
	ISRm19	IS110	1,224	[62, 69]
	ISRm21	ISAs1	>425	[61]
	ISRm23	-	2,598	[61, 65]
	ISRm28	-	>1,196, >1,599	[61, 65]
	ISRm31	IS66	2,803	[70]
	ISRm102F34-1	-	1,481	[32, 33]
	ISRm220-12-3 (ISRm220-13-1)	ISNCY	1,100	[33]

Table 1: cont….

	ISRm220-13-2	-	1,500	[33]
	ISRm220-13-5	-	1,550	[32, 33]
	ISRm2011-2	IS630-Tc1	1,053	[33, 40, 67, 71]
	ISRm2011-3	-	2,000	[33]
	ISSmeSM11a-1	-	1,119	[61]
	ISSmeSM11a-2	-	1,426	[61, 65]
	ISSmeSM11a-3	-	1,835	[61]
	ISSmeSM11a-4	-	1,569	[61]
	ISSmeSM11b	-	2,435	[65]
	syrM-nodD3 intergenic region	-	>885	[72]
Mesorhizobium loti	msi300	-	>357	[73]
	msi301	-	>1,530	[73]
	msi435	-	1,066	[73]
	msi436	-	1,066	[73]
	msi437	-	1,063	[73]
	msi442	-	1,670	[73]
Rhizobium etli	ISRel1	IS4	1,273	[27]
	ISRel2	IS481	1,047	[27]
	ISRel3	IS21	2,308	[27]
	ISRel4	ISAs1	-	[27]
	ISRel5	IS21	2,587	[27]
	ISRel6	IS630	1,047	[27]
	ISRel7	IS66	-	[27]
	ISRel8	ISNCY	-	[27]
	ISRel9	IS5	1,571	[27]
	ISRel10	IS481	-	[27]
	ISRel11	IS5	862	[27]
	ISRel12	IS3	-	[27]
	ISRel14	IS5	-	[27]
	ISRel15	IS66	2,497	[27]
	ISRel16	IS5	-	[27]
	ISRel17	IS3	-	[27]
	ISRel19	IS66	2,798	[27]
	ISRel20	IS110	-	[27]

Table 1: cont....

	ISRel21	ISNCY	-	[27]
	ISRel22	IS1111	-	[27]
	ISRel23	IS1111	-	[27]
	ISRel25	IS630	-	[27]
	ISRel26	IS66	-	[27]
	ISRl2 homolog	IS5	932	[26]
Rhizobium galegae	ISRl2 homolog	IS5	932	[26]
Rhizobium gallicum	ISRel1	IS4	1,273	[27]
	ISRel2	IS481	1,047	[27]
	ISRel3	IS21	2,308	[27]
	ISRel4	ISAs1	-	[27]
	ISRel5	IS21	2,587	[27]
	ISRel8	ISNCY	-	[27]
	ISRel9	IS5	1,571	[27]
	ISRel12	IS3	-	[27]
	ISRel15	IS66	2,497	[27]
	ISRel16	IS5	-	[27]
	ISRel17	IS3	-	[27]
	ISRel19	IS66	2,798	[27]
	ISRel20	IS110	-	[27]
	ISRel21	ISNCY	-	[27]
	ISRel22	IS1111	-	[27]
	ISRel23	IS1111	-	[27]
	ISRel25	IS630	-	[27]
Rhizobium giardinii	ISRel2	IS481	1,047	[27]
	ISRel4	ISAs1	-	[27]
	ISRel9	IS5	1,571	[27]
Rhizobium leguminosarum	ISRel2	IS481	1,047	[27]
	ISRel4	ISAs1	-	[27]
	ISRel8	ISNCY	-	[27]
	ISRel9	IS5	1,571	[27]
	ISRl1	IS66	2,495	[74]
	ISRl2	IS5	932	[26]
	ISRle1	IS110	1,409	[75]
	ISRle39A	IS6	893	[76]
	ISRle39B	IS6	890	[76]

Table 1: cont….

	ISRm2 homolog	IS66	-	[36]
	ISRm3 homolog	IS256	825; 1,145	[37, 38, 77]
	ISRm4 homolog	IS5	900	[37]
	ISRl2 homolog	IS5	932	[26]
	ISRm18 homolog	-	1,015	[77]
	ISRm102F34-1	-	1,481	[32]
	ISRm2011-2 homolog	IS630	441	[77]
	ISR1 homolog	IS3	267	[77]
	IST2 homolog	IS256	329	[77]
	IS511 homolog	IS3	873; 930	[77]
	IS801 homolog	IS91	1,176; 1,197	[77]
	IS869 homolog	IS5	367; 375	[77]
	IS1111A	IS110	1,035	[77]
	IS1312 homolog	IS3	351; 381; 386; 389; 432	[77]
Rhizobium lupini	ISR1	IS3	1,260	[53, 78]
Rhizobium tropici	ISRl2 homolog	IS5	932	[26]
	ISRtr1	IS256	1,364	[49, 79]
	ISRtr2	IS3	1,321	[49]
	ISRtr3	IS5	933	[49]
	ISRtr4	IS5	932	[49]
	ISRtr5	IS166	2,699	[49]
Rhizobium spp.	IS679	IS66	2,704	[57]
	ISRl2 homolog	IS5	932	[26]
	ISRsp3	IS21	2,489	[80]
	ISRLd TAL1145-1	ISNCY	2,500	[45]

bp: base pairs; -: not reported
IS families were assigned on the basis of structural similarities and homology of transposase genes. Members of the IS21 family are hierarchically included in the IS3 family [81].

REFERENCES

[1]   Downie JA, Hombrecher G, Ma QS, *et al.* Cloned nodulation genes of *Rhizobium leguminosarum* determine host-range specificity. Mol Gen Genet 1983; 190: 359-65.

[2]   Sambrook J, Fritsch EF, Maniatis T. Molecular Cloning: A Laboratory Manual. Cold Spring Harbor, New York, USA. 1989.

[3]   Laguerre G, Geniaux E, Mazurier SI, *et al.* Conformity and diversity among field isolates of *Rhizobium leguminosarum* bv. *viciae*, bv. *trifolii*, and bv. *phaseoli* revealed by DNA hybridization using chromosome and plasmid probes. Can J Microbiol 1992; 39: 412-19.

[4]   Masterson RV, Prakash RK, Atherly AG. Conservation of symbiotic nitrogen fixation gene sequences in *Rhizobium japonicum* and *Bradyrhizobium japonicum*. J Bacteriol 1985; 163: 21-26.

[5]   Sullivan JT, Eardly BD, van Berkum P, Ronson CW. Four unnamed species of nonsymbiotic rhizobia isolated from the rhizosphere of *Lotus corniculatus*. Appl Envir Microbiol 1996; 62: 2818-25.

[6]   Minamisawa K, Seki T, Onodera S, *et al.* Genetic relatedness of *Bradyrhizobium japonicum* field isolates as revealed by repeated sequence and various other characteristics. Appl Envir Microbiol 1992; 58: 2832-39.

[7]   Sadowsky MJ, Bohlool BB, Keyser KK. Serological relatedness of *Rhizobium fredii* to other rhizobia and to the bradyrhizobia. Appl Envir Microbiol 1987; 53: 1785-89.

[8]   Yokoyama T, Ando S, Murakami T, Imai H. Genetic variability of the common *nod* gene in soybean bradyrhizobia isolated in Thailand and Japan. Can J Microbiol 1996; 42: 1209-18.

[9]   Appelbaum E. The *Rhizobium/Bradyrhizobium*-Legume Symbiosis. In: Gresshoff M, Ed. Molecular Biology of Symbiotic Nitrogen Fixation. Boca Raton, CRC Press Inc., 1990; pp. 131-58.

[10]   Fischer HM. Genetic regulation of nitrogen fixation in rhizobia. Microbiol Rev 1994; 58:352-86.

[11]   Saito A, Mitsui H, Hattori R, *et al.* Slow-growing and oligotrophic soil bacteria phylogenetically close to *Bradyrhizobium japonicum*. FEMS Microbiol Ecol 1998; 25:277-86.

[12]   Sanjuan J, Carlson RW, Spaink HP, *et al.* A 2-O-methylfucose moiety is present in the lipo-oligosaccharide nodulation signal of *Bradyrhizobium japonicum*. Proc Natl Acad Sci USA. 1992; 89: 8789-93.

[13]   Demezas DH, Reardon TB, Watson JM, Gibson AH. Genetic diversity among *Rhizobium leguminosarum* bv. trifolii strains revealed by allozyme and restriction fragment length polymorphism analyses. Appl Envir Microbiol 1991; 57: 3489-95.

[14]   Wallington EJ, Lund PA. *Rhizobium leguminosarum* contains multiple chaperonin (*cpn60*) genes. Microbiol 1994; 140: 113-22.

[15]   Paffetti D, Scotti C, Gnocchi S, *et al.* Genetic diversity of an Italian *Rhizobium meliloti* population from different *Medicago sativa* varieties. Appl Envir Microbiol 1996; 62: 2279-85.

[16]   Aguilar OM, Lopez MV, Riccillo PM, *et al.* Prevalence of the *Rhizobium etli*-like allele in genes coding for 16S rRNA among the indigenous rhizobial populations found associated with wild beans from the Southern Andes in Argentina. Appl Envir Microbiol 1998; 64: 3520-24.

[17]   Wernegreen JJ, Harding EE, Riley MA. *Rhizobium* gone native: unexpected plasmid stability of indigenous *Rhizobium leguminosarum*. Proc Natl Acad Sci USA. 1997; 94: 5483-88.

[18]   Flores M, Mavingui P, Girard L, *et al.* Three replicons of *Rhizobium* sp. strain NGR234 harbor symbiotic gene sequences. J Bacteriol 1998; 180: 6052-53.

[19]   Zhang XX, Kosier B, Priefer UB. Symbiotic plasmid rearrangement in *Rhizobium leguminosarum* bv. viciae VF39SM. J Bacteriol 2001; 183: 2141-44.

[20]   Silva C, Vinuesa P, Eguiarte LE, *et al. Rhizobium etli* and *Rhizobium gallicum* nodulate common bean (*Phaseolus vulgaris*) in a traditionally managed Milpa Plot in Mexico:

population genetics and biogeographic implications. Appl Envir Microbiol 2003; 69: 884-93.

[21] Gu CT, Wang ET, Tian CF, *et al. Rhizobium miluonense* sp. nov., a symbiotic bacterium isolated from *Lespedeza* root nodules. Int J Syst Evol Microbiol 2008; 58: 1365-68.

[22] Bromfield ESP, Tambong JT, Cloutier S, *et al. Ensifer, Phyllobacterium* and *Rhizobium* species occupy nodules of *Medicago sativa* (alfalfa) and *Melilotus alba* (sweet clover) grown at a Canadian site without a history of cultivation. Microbiol 2010; 156: 505-20.

[23] Tas E, Saano A, Leinonen P, Lindstrom K. Identification of *Rhizobium* spp. in peat-based inoculants by DNA hybridization and PCR and its application in inoculant quality control. Appl Envir Microbiol 1995; 61: 1822-27.

[24] Bjourson AJ, Cooper JE. Isolation of *Rhizobium loti* strain-specific DNA sequences by subtraction hybridization. Appl Envir Microbiol 1988; 54: 2852-55.

[25] Streit W, Bjourson AJ, Cooper JE, Werner D. Application of subtraction hybridization for the development of a *Rhizobium leguminosarum* biovar *phaseoli* and *Rhizobium tropici* group-specific DNA probe. FEMS Microbiol Ecol 1993; 13: 59-67.

[26] Mazurier SI, Rigottier-Gois L, Amarger N. Characterization, distribution, and localization of IS*Rl2*, an insertion sequence element isolated from *Rhizobium leguminosarum* bv. viciae. Appl Envir Microbiol 1996; 62: 685-93.

[27] Lozano L, Hernandez-Gonzalez I, Bustos P, *et al.* Evolutionary dynamics of insertion sequences in relation to the evolutionary histories of the chromosome and symbiotic plasmid genes of *Rhizobium etli* populations. Appl Envir Microbiol 2010; 76: 6504-13.

[28] Krishnan HB, Pueppke SG. Characterization of RFRS9, a second member of the *Rhizobium fredii* repetitive sequence family from the nitrogen-fixing symbiont *R. fredii* USDA 257. Appl Envir Microbiol 1993; 59: 150-55.

[29] Freiberg C, Fellay R, Bairoch A, *et al.* Molecular basis of symbiosis between *Rhizobium* and legumes. Nature 1997; 387: 394-01.

[30] Laberge S, Middleton AT, Wheatcroft R. Characterization, nucleotide sequence, and conserved genomic locations of insertion sequence IS*Rm5* in *Rhizobium meliloti*. J Bacteriol 1995; 177: 3133-42.

[31] Martinez E, Romero D, Palacios R. The *Rhizobium* genome. Crit Rev Plant Sci 1990; 9: 59-93.

[32] Selbitschka W, Zekri S, Schroder G, *et al.* The *Sinorhizobium meliloti* insertion sequence (IS) elements IS*Rm102F34–1*/IS*Rm7* and IS*Rm220-13-5* belong to a new family of insertion sequence elements. FEMS Microbiol Lett 1999; 172: 1-7.

[33] Simon R, Hotte B, Klauke B, Kosier B. Isolation and characterization of insertion sequence elements from gram-negative bacteria by using new broad-host-range, positive selection vectors. J Bacteriol 1991; 173: 1502-08.

[34] Ruvkun GB, Meade HM, van Den Bos RC, Ausubel FM. IS*Rm1*: a *Rhizobium meliloti* insertion sequence that transposes preferentially into nitrogen fixation genes. J Mol Appl Gen 1982; 1: 405-18.

[35] Wheatcroft R, Watson RJ. Distribution of insertion sequence IS*Rm1* in *Rhizobium meliloti* and other gram-negative bacteria. J Gen Microbiol 1988; 134: 113-21.

[36] Dusha I, Kovalenko S, Banfalvi Z, Kondorosi A. *Rhizobium meliloti* insertion element IS*Rm2* and its use for identification of the *fixX* gene. J Bacteriol 1987; 169: 1403-09.

[37] Villadas PJ, Burgos P, Rodriguez-Navarro DN, *et al.* Characterization of rhizobia homologues of *Sinorhizobium meliloti* insertion sequences IS*Rm3* and IS*Rm4*. FEMS Microbiol Ecol 1998; 25: 341-48.

[38] Wheatcroft R, Laberge S. Identification and nucleotide sequence of *Rhizobium meliloti* insertion sequence IS*Rm3*: similarity between the putative transposase encoded by IS*Rm3* and those encoded by *Staphylococcus aureus* IS*256* and *Thiobacillus ferrooxidans* IS*T2*. J Bacteriol 1991; 173: 2530-38.

[39] Andronov EE, Roumiantseva ML, Simarov BV. Genetic diversity of a natural population of *Sinorhizobium meliloti* revealed in analysis of cryptic plasmids and IS*Rm2011-2* fingerprints. Russ J Genet 2001; 37:494-99.

[40] Segundo E, Martinez-Abarca F, van Dillewijn P, *et al.* Characterisation of symbiotically efficient alfalfa-nodulating rhizobia isolated from acid soils of Argentina and Uruguay. FEMS Microbiol Ecol 1999; 28: 169-76.

[41] Gonzalez V, Santamaria R, Bustos P, *et al.* The partitioned *Rhizobium etli* genome: genetic and metabolic redundancy in seven interacting replicons. Proc Natl Acad Sci USA. 2006; 103: 3834-39.

[42] Kaluza K, Hahn M, Hennecke H. Repeated sequences similar to insertion elements clustered around the *nif* region of the *Rhizobium japonicum* genome. J Bacteriol 1985; 162: 535-542.

[43] Minamisawa K, Isawa T, Nakatsuka Y, Ichikawa N. New *Bradyrhizobium japonicum* strains that possess high copy numbers of the repeated sequence RSα. Appl Envir Microbiol 1998; 65: 1845-51.

[44] Barran LR, Bromfield ESP, Laberge S, Wheatcroft R. Insertion sequence (IS) hybridization supports classification of *Rhizobium meliloti* by phage typing. Mol Ecol 1994; 3: 267-70.

[45] Rice DJ, Somasegaran P, MacGlashan K, Bohlool BB. Isolation of insertion sequence IS*RLd*TAL1145-1 from a *Rhizobium* sp. (*Leucaena diversifolia*) and distribution of homologous sequences identifying cross-inoculation group relationships. Appl Envir Microbiol 1994; 60: 4394-03.

[46] Niemann S, Puhler A, Tichy HV, *et al.* Evaluation of the resolving power of three different DNA fingerprinting methods to discriminate among isolates of a natural *Rhizobium meliloti* population. J Appl Microbiol 1997; 82: 477-84.

[47] Andronov EE, Terefework Z, Roumiantseva ML, *et al.* Symbiotic and genetic diversity of *Rhizobium galegae* isolates collected from the *Galega orientalis* gene center in the Caucasus. Appl Envir Microbiol 2003; 69: 1067-74.

[48] Deng W, Gordon MP, Nester EW. Sequence and distribution of IS*1312*: evidence for horizontal DNA transfer from *Rhizobium meliloti* to *Agrobacterium tumefaciens*. J Bacteriol 1995; 177: 2554-59.

[49] Hernandez-Lucas I, Ramirez-Trujillo JA, Gaitan MA, *et al.* Isolation and characterization of functional insertion sequences of rhizobia. FEMS Microbiol Lett 2006; 261: 25-31.

[50] Wheatcroft R, Watson RJ. A positive strain identification method for *Rhizobium meliloti*. Appl Envir Microbiol 1988; 54: 574-76.

[51] Judd AK, Sadowsky MJ. The *Bradyrhizobium japonicum* serocluster 123 hyperreiterated DNA region, HRS1, has DNA and amino acid sequence homology to IS*1380*, an insertion sequence from *Acetobacter pasteurianus*. Appl Envir Microbiol 1993; 59: 1656-61.

[52] Isawa T, Sameshima R, Mitsui H, Minamisawa K. IS*1631* occurrence in *Bradyrhizobium japonicum* highly reiterated sequence-possessing strains with high copy numbers of repeated sequences RSα and RSβ. Appl Envir Microbiol 1999; 65: 3493-01.

[53] Hahn M, Hennecke H. Mapping of a *Bradyrhizobium japonicum* DNA region carrying genes for symbiosis and an asymmetric accumulation of reiterated sequences. Appl Envir Microbiol 1987; 53: 2247-52.

[54]   Priefer UB, Burkardt HJ, Klipp W, Phuler A. IS*R1:* an insertion element isolated from the soil bacterium *Rhizobium lupini.* Cold Spring Harb Symp Quant Biol 1981; 45: 87-91.

[55]   Chaintreuil C, Boivin C, Dreyfus B, Giraud E. Characterization of the common nodulation genes of the photosynthetic *Bradyrhizobium* sp. ORS285 reveals the presence of a new insertion sequence upstream of *nodA.* FEMS Microbiol Lett 2001; 194: 83-86.

[56]   Vinardell JM, Ollero GJ, Krishnan HB, *et al.* IS*Rf1*, a transposable insertion sequence from *Sinorhizobium fredii.* Gene 1997; 204: 63-69.

[57]   Han CG, Shiga Y, Tobe T, *et al.* Structural and functional characterization of IS*679* and IS*66*-family elements. J Bactriol 2001; 183: 4296-04.

[58]   Soto MJ, Zorzano A, Olivares J, Toro N. Nucleotide sequence of *Rhizobium meliloti* GR4 insertion sequence IS*Rm3* linked to the nodulation competitiveness locus *nfe.* Plant Mol Biol 1992; 20: 307-09.

[59]   Soto MJ, Zorzano A, Olivares J, Toro N. Sequence of IS*Rm4* from *Rhizobium meliloti* strain GR4. Gene 1992; 120: 125-26.

[60]   Zekri S, Soto MJ, Toro N. IS*Rm4-1* and IS*Rm9*, two novel insertion sequences from *Sinorhizobium meliloti.* Gene 1998; 207: 93-96.

[61]   Stiens M, Schneiker S, Keller M, *et al.* Sequence analysis of the 144-kilobase accessory plasmid pSmeSM11a, isolated from a dominant *Sinorhizobium meliloti* strain identified during a long-term field release experiment. Appl Envir Microbiol 2006; 72: 3662-72.

[62]   Tobes R, Pareja E. Bacterial repetitive extragenic palindromic sequences are DNA targets for insertion sequence elements. BMC Genomics 2006; 7: 62.

[63]   Zekri S, Toro N. Identification and nucleotide sequence of *Rhizobium meliloti* insertion sequence IS*Rm6,* a small transposable element that belongs to the IS*3* family. Gene 1996; 175: 43-48.

[64]   Zekri S, Toro N. A new insertion sequence from *Sinorhizobium meliloti* with homology to IS*1357* from *Methylobacterium* sp. and IS*1452* from *Acetobacter pasteurianus.* FEMS Microbiol Lett 1998; 158: 83-87.

[65]   Stiens M, Schneiker S, Puhler A, Schluter A. Sequence analysis of the 181-kb accessory plasmid pSmeSM11b, isolated from a dominant *Sinorhizobium meliloti* strain identified during a long-term field release experiment. FEMS Microbiol Lett 2007; 271: 297-09.

[66]   Biondi EG, Fancelli S, Bazzicalupo M. IS*Rm10*: a new insertion sequence of *Sinorhizobium meliloti*: nucleotide sequence and geographic distribution. FEMS Microbiol Lett 1999; 181: 171-76.

[67]   Martinez-Abarca F, Toro N. RecA-independent ectopic transposition *in vivo* of a bacterial group II intron. Nucleic Acids Res 2000; 28: 4397-02.

[68]   Schneiker S, Kosier B, Puhler A, Selbitschka W. The *Sinorhizobium meliloti* insertion sequence (IS) element IS*Rm14* is related to a previously unrecognized IS element located adjacent to the *Escherichia coli* locus of enterocyte effacement (LEE) pathogenicity island. Curr Microbiol 1999; 39: 274-81.

[69]   Choi S, Ohta S, Ohtsubo E. A novel IS element, IS*621*, of the IS*110*/IS*492* family transposes to a specific site in repetitive extragenic palindromic sequences in *Escherichia coli.* J Bacteriol 2003; 185: 4891-00.

[70]   Biondi EG, Femia AP, Favilli F, Bazzicalupo M. IS*Rm31*, a new insertion sequence of the IS*66* family in *Sinorhizobium meliloti.* Arch Microbiol 2003; 180: 118-26.

[71]   Selbitschka W, Arnold W, Jording D, *et al.* The insertion sequence element IS*Rm2011-2* belongs to the IS*630*-Tc1 family of transposable elements and is abundant in *Rhizobium meliloti.* Gene 1995; 163: 59-64.

[72]   Barnett MJ, Rushing BG, Fisher RF, Long SR. Transcription start sites for *syrM* and *nodD3* flank an insertion sequence relic in *Rhizobium meliloti*. J Bacteriol 1996; 178: 1782-87

[73]   Sullivan JT, Trzebiatowski JR, Cruickshank RW, *et al.* Comparative sequence analysis of the symbiosis island of *Mesorhizobium loti* strain R7A. J. Bacteriol 2002; 184: 3086-95.

[74]   Ponsonnet C, Normand P, Pilate G, Nesme X. IS*292*: a novel insertion element from *Agrobacterium*. Microbiol 1995; 141: 853-61.

[75]   Partridge SR, Hall RM. The IS*1111* family members IS*4321* and IS*5075* have subterminal inverted repeats and target the terminal inverted repeats of Tn*21* family transposons. J Bacteriol 2003; 85: 6371-84.

[76]   Rochepeau P, Selinger LB, Hynes MF. Transposon-like structure of a new plasmid-encoded restriction-modification system in *Rhizobium leguminosarum* VF39SM. Mol Gen Genet 1997; 256: 387-96.

[77]   Young JP, Crossman LC, Johnston AW, *et al.* The genome of *Rhizobium leguminosarum* has recognizable core and accessory components. Genome Biol 2006; 7: R34.

[78]   Priefer UB, Kalinowski J, Ruger B, *et al.* IS*R1*, a transposable DNA sequence resident in *Rhizobium* class IV strains, shows structural characteristics of classical insertion elements. Plasmid 1989; 21: 120-28.

[79]   Mavingui P, Laeremans T, Flores M, *et al.* Genes essential for Nod factor production and nodulation are located on a symbiotic amplicon (AMPRtrCFN299pc60) in *Rhizobium tropici*. J Bacteriol 1998; 180: 2866-74.

[80]   Hashimoto M, Fukui M, Hayano K, Hayatsu M. Nucleotide sequence and genetic structure of a novel carbaryl hydrolase gene (*cehA*) from *Rhizobium* sp. strain AC100. Appl Envir Microbiol 2002; 68: 1220-27.

[81]   Ohtsubo E, Sekine Y. Bacterial insertion sequences. Curr Top Microbiol Immunol 1996; 204: 1-26.

CHAPTER 8

Genotypic Diversity of Rhizobia Assessed by Sequence Analysis

Neelawan Pongsilp[*]

Department of Microbiology, Faculty of Science, Silpakorn University, Nakhon Pathom, Thailand

Abstract: The identification of bacteria based on phenotypic characteristics is generally not accurate because several species are difficult to be distinguished phenotypically. In case of rhizobia, the current classification is mainly based on DNA sequences [especially a DNA sequence encoding small subunit ribosomal RNA (16S rDNA)], DNA homologies, phylogenetic relationships and the locations of symbiotic genes. The 16S rDNA is very useful for estimating the evolutionary relationships and identifying bacteria. The most dramatic progress in the construction of microbial phylogeny is based on sequence analysis of 16S rDNA. The 16S rDNA sequences mainly support the proposal of novel genera and species of rhizobia. In some cases, several genera are identical in 16S rDNA sequence analysis. The other regions, such as large subunit ribosomal RNA gene (23S rDNA) as well as intergenic spacer between 16S and 23S rRNA sequences (16S-23S IGS), are suitable alternatives for classification and identification purposes. Multilocus sequence analysis (MLSA), which employs a set of nucleotide sequences including 16S rDNA, house keeping genes and symbiotic genes, has the greater potential for rhizobial classification. Sequence analysis is currently the most promising method for identification of rhizobial genera.

Keywords: Dendrogram, Genotypic diversity, Housekeeping gene, Intergenic spacer between 16S and 23S rRNA sequences (16S-23S IGS), Multilocus sequence analysis (MLSA), Multilocus sequence typing (MLST), Phylogenetic tree, Rhizobia, Small subunit ribosomal RNA (16S rDNA), Symbiotic gene.

8.1. SMALL SUBUNIT RIBOSOMAL RNA GENE (16S rDNA) SEQUENCE ANALYSIS OF RHIZOBIA

8.1.1. Principle of 16S rDNA Sequence Analysis

The 16S rRNA gene, also designated "16S rDNA", is a DNA sequence encoding small subunit ribosomal RNA. This sequence is about 1.5 kilobases (kb) long and is composed of both highly conserved and variable regions. The variable region of

*Address correspondence to Neelawan Pongsilp: Department of Microbiology, Faculty of Science, Silpakorn University, Nakhon Pathom, Thailand; Tel: +66-34-245337; Fax: +66-34-245336-37; E-mail: neelawan@su.ac.th

16S rDNA can be amplified in the polymerase chain reaction (PCR) using primers flanking this region and the sequence of the variable region is used for the comparative taxonomy [1]. Pairs of universal primers (forward primers and reverse primers) and PCR conditions were designed for the amplification of 16S rDNA in most eubacteria, as shown in Table **1**. The generated DNA sequence is assembled by aligning with sequences in database libraries such as Genbank sequence database of the National Center for Biotechnology Information (NCBI) (www.ncbi.nlm.nih.gov/), the Ribosomal Database Project (RDP) (http://rdp.cme.msu.edu/) and the European Molecular Biology Laboratory (EMBL) nucleotide sequence database (www.ebi.ac.uk/embl/). Direct sequencing of 16S rDNA has been used to establish the genetic relationships and to characterize strains at species or higher levels. The 16S rDNA is useful for estimating the evolutionary relationships among bacteria because it is slowly evolving and the gene product is both universally essential and functionally conserved [2]. The most dramatic progress in the construction of microbial phylogeny is based on sequence analysis of 16S rDNA. The accuracy of 16S rDNA sequencing in the identification of bacteria has been proved superior over phenotypic methods. This technique is discriminatory and reproducible. Sequence comparisons of 16S rDNA have become the standard method for assessing phylogenetic relationships among bacteria [3]. However, this region is not appropriate to differentiate strains within species [4]. The quality of sequence is an important feature needed to be considered. The good quality of sequence is presented as no ambiguous bases (multiple Ns or more than one base at a particular position). The phylogenetic tree for the DNA sequences can be constructed using the Neighbour-Joining (NJ) method [5]. Sequences are taken together in the calculations of levels of sequence similarity using CLUSTALW 1.74 [6] or CLUSTALX softwares [7]. The evolutionary distances are computed using the Maximum Composite Likelihood method [8] and are in units of the number of base substitutions per site. All positions containing alignment gaps and missing are eliminated only in pairwise sequence comparisons. The phylogenetic tree is conducted in the MEGA 3.1 [9] or the MEGA 4 programs [10].

The use of 16S rDNA sequencing has been instrumental in the discrimination of rhizobia in recent years. The full-length sequence analysis of 16S rDNA is one of the most important methods to estimate the phylogeny of rhizobia [11], while the partial 16S rDNA sequencing has been used for rapid screening of the phylogenetic relationships among a large number of rhizobia. In general, a 500-base pairs (bp) sequence can provide adequate differentiation for the identification of rhizobia.

8.1.2. Phylogenetic Analysis Constructed from 16S rDNA Sequences of Rhizobial Genera

Novel members of rhizobia in genera *Burkholderia*, *Cupriavidus/Ralstonia*, *Devosia*, *Herbaspirillum*, *Methylobacterium*, *Microvirga*, *Ochrobactrum* and *Phyllobacterium* have been discovered by 16S rDNA sequence analyses [12-21]. These findings suggest that the genes responsible for symbiosis with legumes are transmissible horizontally and function in a relatively wide range of bacterial taxa [17, 22]. Phylogenetic analyses of 16S rDNA have been constructed in the previous studies. According to Ngom *et al.* [16], the clusters in the phylogenetic tree, which was constructed based on the nearly full-length 16S rDNA, correlated well with the taxonomy of strains: i) the first cluster contained *Blastobacter* (currently *Bradyrhizobium*) and *Bradyrhizobium* in the family Bradyrhizobiaceae; ii) the second cluster contained *Ochrobactrum* in the family Brucellaceae; iii) the third cluster consisted of 2 genera *Mesorhizobium* and *Phyllobacterium* in the family Phyllobacteriaceae; iv) the fourth cluster consisted of genera *Allorhizobium* (currently *Rhizobium*), *Rhizobium* and *Sinorhizobium* in the family Rhizobiaceae.

There are the known instances in which 16S rDNA sequencing cannot differentiate among a limited number of genera. Even though *Rhizobium* is the genus of nitrogen-fixing symbiont and *Agrobacterium* is the genus of plant pathogen, these 2 genera are phylogenetically entwined with one another and cannot be separated by their 16S rDNA sequences [23]. The main criteria to distinguish *Rhizobium* from *Agrobacterium* base on morphological characteristics and the nodule-producing ability of *Rhizobium* or the tumor-producing ability of *Agrobacterium tumefaciens* or the ability to induce hairy root of *Agrobacterium rhizogenes*. Young *et al.* [24] proposed that all *Agrobacterium* species were more properly considered to be members of *Rhizobium*. Based on 16S rDNA sequences, the genus *Bradyrhizobium* has been classified into a clade in the Proteobacteria along with oligotrophic soil or aquatic bacteria such as *Nitrobacter winogradskyi*, *Rhodoplanes roseus* and *Rhodopseudomonas palustris* as well as the human pathogen *Afipia* spp. [21, 25-27]. Analysis of the full-length 16S rDNA sequences indicated that *Agromonas oligotrophica* was more closely related to *Bradyrhizobium japonicum* (similarity values: 98.1% to 98.8%) than the other strains such as *Afipia* spp., *Bradyrhizobium elkanii*, *Nitrobacter* spp. and *Rho. palustris* [25]. The full-length or nearly full-length 16S rDNA sequences of rhizobia and some related species are aligned, and percentage homologies are shown in Table **2**. As displayed in Fig. **1**, a phylogenetic tree based on the 16S rDNA sequences was constructed using the NJ method [5]. Sequences were taken together in the calculations of levels of sequence similarity using CLUSTALW 1.74 [6]. The percentage of replicate trees in which the associated taxa clustered together in the bootstrap test (1,000 replicates) is shown next to the branches. The evolutionary distances were computed using the Maximum Composite Likelihood method [8] and are in units of the number of base substitutions per site. All positions containing alignment gaps and missing data were eliminated only in pair wise sequence comparisons. The phylogenetic tree was conducted in the MEGA4 suite of program [10]. Rhizobia in 13 genera formed 2 main clusters. Members within the first cluster belonged to the alpha (α) subclass of Proteobacteria. This cluster was composed of 10 genera including *Azorhizobium*, *Bradyrhizobium*, *Devosia*, *Ensifer* (or *Sinorhizobium*), *Mesorhizobium*, *Methylobacterium*, *Ochrobactrum*, *Phyllobacterium*, *Rhizobium* and *Shinella*. A dendrogram showed that *Rhizobium* and *Ensifer* (or *Sinorhizobium*) were closely related to each other. *Shinella* was the closest neighbour of *Mesorhizobium*. Members within the second cluster belonged to the beta (β) subclass of Proteobacteria. *Cupriavidus* (or *Ralstonia*) and *Burkholderia* were more closely related to each other than to *Herbaspirillum*.

Table 1: Primer pairs used in the PCR amplification of 16S rDNA

Primer Pair and Sequence (5' to 3')	PCR Mixture	PCR Cycling Condition	PCR Product Size (bp)	References
Forward: fD1 (AGA GTT TGA TCC TGG CTC AG) Reverse: rD1 (AAG GAG GTG ATC	1 ng/μl template DNA; 1X buffer; 1.5 mM MgCl$_2$; 5% dimethyl sulfoxide; 200 μM dNTPs;	95°C 3 min; 35 cycles x (94°C 70 sec, 56°C 40 sec and 72°C 130 sec); 72°C 370 sec	1,500	[28]

Table 1: cont…

CAG CC)	0.3 μM each primer; 0.02 unit/μl *Taq* DNA polymerase			
Forward: 16Sa (CGC TGG CGG CAG GCT TAA CA) Reverse: 16Sb (CCA GCC GCA GGT TCC CCT)	2 ng/μl template DNA; 1X buffer; 100 mM MgCl₂; 10 mM dNTPs; 0.08 μM each primer; 0.025 unit/μl *Taq* DNA polymerase	95°C 5 min; 35 cycles x (94°C 1 min, 55°C 1 min and 72°C 2 min); 72°C 3 min	1,500	[29]
Forward: f (AGA GTT TGA TCC TGG CTC AG) Reverse: r (AAG GAG GTG ATT CCA GCC)	2 ng/μl template DNA; 2.5 mM 10X buffer; 0.4 nM dNTPs; 0.8 μM each primer; 0.04 unit/μl *Taq* DNA polymerase	94°C 15 min; 34 cycles x (94°C 1 min, 55°C 1 min and 72°C 1 min); 72°C 10 min	1,350	[30]
Forward: PF2 (TAC TGT CGA TCT GGA GTA TG) Reverse: PR1 (ATT GTA GCA CGT GTG TAG CC)	template DNA; 1X buffer; 1.5 mM dNTPs; 0.2 μM each primer; 0.04 unit/μl *Taq* DNA polymerase	94°C 3 min; 30 cycles x (94°C 1 min, 60°C 1 min and 72°C 1 min); 72°C 5 min	558	[31]
Forward: UN16S 926f (AAA CTY AAA KGA ATT GAC GG) Reverse: UN16S 1392r (ACG GGC GGT GTG TRC)	2 ng/μl template DNA; 1X buffer; 1.5 mM MgCl₂; 0.2 mM dNTPs; 2 μM each primer; 0.05 unit/μl *Taq* DNA polymerase	98°C 5 min; 34 cycles x (95°C 30 sec, 62°C 30 sec and 72°C 1 min); 72°C 5 min	500	[29]
Forward: F984 (AAC GCG AAG AAC CTT AC) Reverse: R1378 (CGG TGT GTA CAA GGC CCG GGA ACG)	2 ng/μl template DNA; 1.2X buffer; 2.5 mM MgCl₂; 1.3 mM dNTPs; 0.4 μM each primer; 0.15 unit/μl *Taq* DNA polymerase	95°C 5 min; 35 cycles x (94°C 1 min, 53°C 1 min and 72°C 1 min); 72°C 10 min	430	[32]
Forward: AR1 (GAG AAT TCC TGG CTC AGA ACG AAC GCT GGC G) Reverse: AR2 (CGA AGC TTC CCA CTG CTG CCT CCC GTA GGA)	0.1-0.2 ng/μl template DNA; 2.5 mM MgCl₂; 50 μM dNTPs; 1 nM each primer; 0.0125 unit/μl *Taq* DNA polymerase	93°C 2 min; 34 cycles x (93°C 45 sec, 62°C 45 sec and 72°C 2 min); 72°C 5 min	268	[33]

bp: base pairs.

Mixed bases nomenclature: K, G or T; R, A or G; Y, C or T.

Table 2: Percentage homologies in 16S rDNA sequences of rhizobia and some related species

Species	1	2	3	4	5	6	7	8	9	10	11	12	13	14	15	16	17	18	19	20
1. Rhi. tropici	100	99	98	96	94	95	93	94	91	91	89	89	89	89	89	89	88	78	81	78
2. Agr. rhizogenes*		100	98	96	94	95	93	94	91	91	89	88	89	89	88	89	88	79	81	78
3. Rhi. leguminosarum			100	96	95	95	93	93	91	91	89	89	89	89	89	89	89	82	82	82
4. Ens. fredii				100	95	96	95	96	91	91	90	89	89	89	89	89	88	78	81	81
5. S. kummerowiae					100	95	95	94	91	91	89	89	89	89	88	88	88	80	81	81
6. O. cytisi						100	94	95	92	91	91	90	90	90	91	90	89	79	81	78
7. Mes. loti							100	94	90	91	90	88	88	88	88	88	88	78	80	80
8. P. trifolii								100	92	92	90	88	88	89	88	88	88	78	81	78
9. Azo. caulinodans									100	90	91	90	90	90	90	90	90	78	78	78
10. D. neptuniae										100	89	88	89	89	89	88	88	79	80	78
11. Bra. elkanii											100	97	97	97	97	95	90	78	80	78
12. Bra. japonicum												100	98	98	98	96	90	78	80	77
13. Agromonas sp.*													100	98	98	96	90	78	80	78
14. N. vulgaris*														100	98	97	90	78	80	78
15. Rho. palustris*															100	96	90	79	80	78
16. Afi. felis*																100	90	78	80	78
17. Met. nodulans																	100	78	80	77
18. C. taiwanensis																		100	92	91
19. Bur. mimosarum																			100	91
20. H. lusitanum																				100

*non-rhizobia

1. *Rhizobium tropici* strain Br859 (Genbank accession no. HQ394213) [1,472 base pairs (bp)]
2. *Agrobacterium rhizogenes** strain CU10 (Genbank accession no. EF522124) (1,474 bp)
3. *Rhizobium leguminosarum* strain USDA 2671 (Genbank accession no. EU488755) (1,422 bp)
4. *Ensifer* (formerly *Sinorhizobium*) *fredii* strain USDA 205 (Genbank accession no. NR_036957) (1,437 bp)
5. *Shinella kummerowiae* strain CCBAU 25048 (Genbank accession no. NR_044066) (1,372 bp)
6. *Ochrobactrum cytisi* strain ESC1 (Genbank accession no. NR_043184) (1,476 bp)
7. *Mesorhizobium loti* strain NGT514 (Genbank accession no. AB289614) (1,474 bp)
8. *Phyllobacterium trifolii* strain PETP02 (Genbank accession no. NR_043193) (1,478 bp)
9. *Azorhizobium caulinodans* strain ORS571 (Genbank accession no. NR_036941) (1,467 bp)
10. *Devosia neptuniae* strain J1 (Genbank accession no. AF469072) (1,478 bp)
11. *Bradyrhizobium elkanii* strain SEMIA 5002 (Genbank sccession no. FJ390895) (1,481 bp)
12. *Bradyrhizobium japonicum* strain USDA 110 (Genbank accession no. RHB16SRDB) (1,457 bp)
13. *Agromonas* sp.* strain S80 (Genbank accession no. AB531476) (1,449 bp)
14. *Nitrobacter vulgaris** strain DSM 10236 (Genbank accession no. NR_042449) (1,441 bp)
15. *Rhodopseudomonas palustris** strain TUT3613 (Genbank accession no. AB498821) (1,484 bp)
16. *Afipia felis** (Genbank accession no. NR_044742) (1,425 bp)
17. *Methylobacterium nodulans* strain ORS2060 (Genbank accession no. NR_027539) (1,435 bp)
18. *Cupriavidus* (formerly *Ralstonia*) *taiwanensis* strain LMG 19424 (Genbank accession no. AF300324) (1,542 bp)
19. *Burkholderia mimosarum* strain PAS44 (Genbank accession no. NR_043167) (1,439 bp)
20. *Herbaspirillum lusitanum* (Genbank accession no. AF543312) (1,524 bp)

8.2. LARGE SUBUNIT RIBOSOMAL RNA GENE (23S rDNA) SEQUENCE ANALYSIS OF RHIZOBIA

Although 16S rDNA has been most widely used, 23S rRNA gene, also designated "23S rDNA", encoding large subunit ribosomal RNA has also been studied. The 23S rDNA has the virtues of 16S rDNA but because of its larger size (about 2.3

kb), it contains more information [34]. The 23S rDNA provides several dramatic differences which may be helpful for classification and identification purposes [2]. The 23S rDNA dendrogram always showed deeper branching than did the 16S rDNA dendrogram and more genotypes were resolved, although in some case, the sequence divergence was not particularly high. *Rhizobium tropici* and *Agr. rhizogenes,* which were identical in 16S rDNA sequences, could be differentiated by 23S rDNA [34]. Moreover, it was found that the grouping of strains in the trees based on 16S rDNA sequences and 23S rDNA sequences were generally consistent [35].

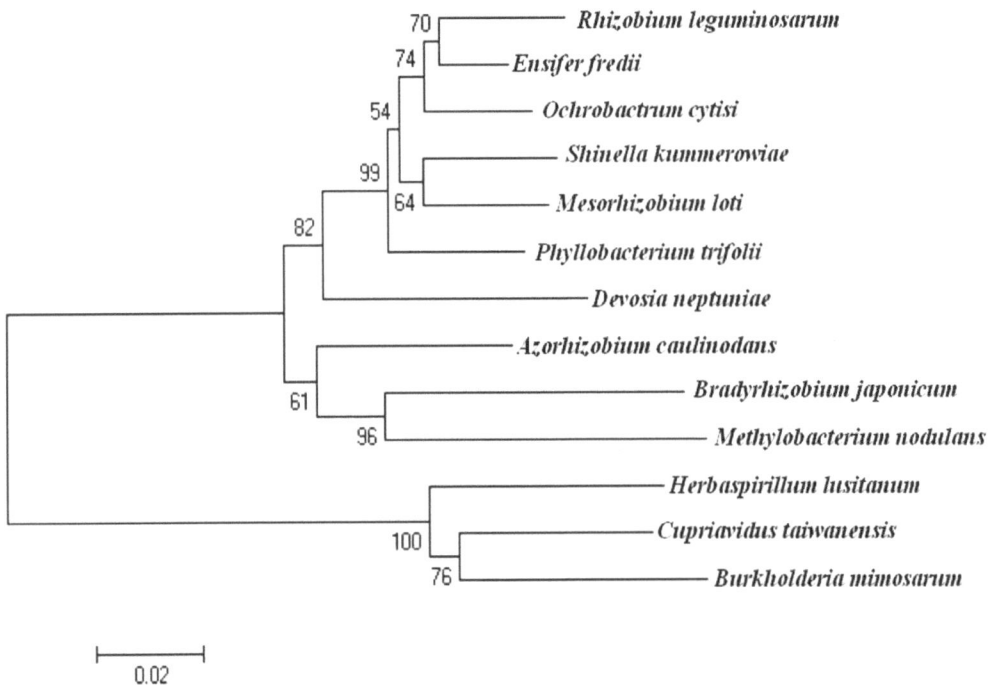

Figure 1: Phylogentic tree based on 16S rDNA sequences.
The numbers at the node are bootstrap values based on 1,000 re-samplings. *Bar* Mutations per sequence position.
Rhi. leguminosarum (Genbank accession no. EU488755); *Ens. fredii* (Genbank accession no. NR_036957); *S. kummerowiae* (Genbank accession no. NR_044066); *O. cytisi* (Genbank accession no. NR_043184); *Mes. loti* (Genbank accession no. AB289614); *P. trifolii* (Genbank accession no. NR_043193); *Azo. caulinodans* (Genbank accession no. NR_036941); *D. neptuniae* (Genbank accession no. AF469072); *Bra. japonicum* (Genbank accession no. RHB16SRDB); *Met. nodulans* (Genbank accession no. NR_027539); *C. taiwanensis* (Genbank accession no. AF300324); *Bur. mimosarum* (Genbank accession no. NR_043167); *H. lusitanum* (Genbank accession no. AF543312).

8.3. INTERGENIC SPACER BETWEEN 16S AND 23S rRNA GENES (16S-23S IGS) SEQUENCE ANALYSIS OF RHIZOBIA

The intergenic spacer between 16S and 23S rRNA sequences, termed as "16S-23S IGS region", has received an increased attention as a target in molecular detection and identification [4, 36]. In contrast to 16S rDNA and 23S rDNA, which are remarkably well conserved throughout most bacterial species, IGS regions exhibit a large degree of sequence diversity and length variation at the levels of genus and species [37]. Even within species, IGS sequence variation may be very high, allowing the differentiation within species [4, 38]. This sequence is very valuable for strain differentiation and identification, making the IGS sequence analysis a very attractive technique for phylogenetic studies of bacteria. The suitability of IGS sequences for the differentiation of closely related rhizobial taxa and for the specific detection of strains in the environment has been assessed in the previous studies. The IGS sequence analyses allowed intra-species differentiation, especially in the genus *Bradyrhizobium*. For the general IGS-PCR, a forward primer, 926f (5' GGT TAA AAC T(C/T)A AA(G/T) GAA TTG ACG G 3'), corresponding to a conserved region of the 3'end of *Escherichia coli* 16S rDNA, sequence positions 901 to 926, and a reverse primer, 115r/23S (5' CCG GGT T(T/G/C)C CCC ATT CGG 3'), corresponding to a conserved region of the 5' end of *Esc. coli* 23S rDNA, sequence positions 97 to 115, were used to amplify the IGS region. Besides the general IGS-PCR, strain-specific primers were developed from the highly variable regions of the specific genera. The primer pair R3ssf (5' GAG CGC TGT GCG ATG CAT CG 3') and R3ssr (5' GCT CAT CTT GCG ATG AAC GAG 3') amplified a 451-bp fragment for *Bradyrhizobium* sp., while the primer pair R2ssf (5' CCT GGA TCA ACG CGG TAT 3') and R2ssr (5' CCA TAG CCG CTC CAA AGG A 3') amplified a 988-bp fragment for *Rhizobium* sp. [39].

8.4. MULTILOCUS SEQUENCE ANALYSIS (MLSA) OF RHIZOBIA

The disadvantage of 16S rDNA phylogeny is that closely related species, or even different genera, cannot always be distinguished because of the high level of sequence conservation. To overcome this limitation, protein-encoding genes with a higher level of sequence divergence than 16S rDNA, but sufficiently conserved to retain phylogenetic signal, have been suggested as alternative phylogenetic markers [40]. Multilocus sequence analysis (MLSA) or multilocus sequence typing (MLST) is a sequence-based method used to characterize bacterial genomes. It has been commonly used to study the population genetic structure and the phylogenetic relatedness within diverse groups of bacteria. In this method, nucleotide sequences of a fixed set of common loci are obtained from a collection of strains and polymorphic sites among these sequences are used to derive an allelic profile or sequence type (ST) for each genome. Comparisons of the resulting data can be used to infer the phylogenetic relationships among bacteria in the sample population [41]. The 16S rDNA, together with housekeeping genes essential for cell maintenance and activity (such as *atpA*, *dnaK*, *gyrB*, *recA* and *rpoB*), are frequently used in MLSA. Metabolite genes are also good alternatives for MLSA. The genes frequently used in MLSA are presented in Table **3**.

MLSA has been used to determine patterns of chromosomal evolutionary descent among rhizobial population. MLSA has been confirmed as a rapid and reliable method for providing information on the phylogenetic relationships and for identifying rhizobial strains representative of novel species [42]. MLSA of housekeeping genes had the greater potential for discrimination of *Ensifer* species than did 16S rDNA. MLSA also supports the proposal to unite *Ensifer* and *Sinorhizobium* in a single genus [40]. Phylogenetic analyses of rhizobia in 5 genera and some non-symbiotic isolates in the alpha (α) subclass of Proteobacteria revealed high congruence between trees constructed from *dnaK* (encoding a heat shock protein, chaperone) and 16S rDNA, but they were not identical. The *dnaK* tree exhibited good resolution in the cases of the genera *Mesorhizobium*, *Rhizobium* and *Sinorhizobium*, even better than usually shown by 16S rDNA phylogeny [43]. By using 10 chromosomal loci, 91 different profiles of MLST were identified among 231 *Medicago*-nodulating rhizobial strains, indicating an extensive diversity in natural populations of *Medicago*-nodulating rhizobia [44]. Based on MLST, 148 isolates that nodulate *Medicago laciniata* and *Medicago truncatula* (barrel medic) were placed into 26 chromosomal sequence types which 99.95% of the isolates had chromosomal genotypes similar to those recovered from *Med. truncatula*. The isolates recovered from *Med. laciniata* were less diverse than those recovered from *Med. truncatula* [45]. MLSA of 4 sequences (*atpD*, *recA*, *glnII* and *rpoB*) could classify 109 *Bradyrhizobium* isolates into 4 species, including *Bra. elkanii*, *Bra. japonicum*, *Bra. liaoningense* and *Bra. yuanmingense*, as well as a novel *Bradyrhizobium* lineage. The suitability of MLSA for making refined ecological and evolutionary inferences has also been demonstrated. *Bra. yuanmingense* was recovered from sites with humid equatorial climates or dry, hot, semiarid climates. While *Bra. elkanii*, *Bra. japonicum* and *Bra. liaoningense* were preferentially recovered from areas with humid, temperate climates with dry winters and hot summers [46].

Symbiotic genes have always been employed in MLSA to examine the genetic structure and the diversity of rhizobia. The symbiotic genes *nifH*, coding for the nitrogenase reductase of the nitrogenase enzyme [47], and *nodC*, coding for the chitin synthase for synthesis of Nod factors [48], are available for comparison. The *nifH* and *nodC* phylogenetic trees represented the variation among the symbiotic genes of different genera including *Bradyrhizobium*, *Rhizobium* and *Sinorhizobium*. Contrary to the 16S rDNA tree in which members of *Rhizobium* and *Sinorhizobium* were placed in different clades, both genera were entwined with one another and placed into the same clade of the *nifH* tree. However, the *nifH* gene of the genus *Bradyrhizobium* was more divergent from the other 2 genera and generated a different cluster, resembling with those shown in the 16S rDNA tree. The *nodC* tree was not congruent with the 16S rDNA tree as members of *Bradyrhizobium*, *Rhizobium* and *Sinorhizobium* were tightly linked and placed in the same branches of the *nodC* tree [49]. Previous studies has also demonstrated that the *nifH* and *nodC* phylogenies are generally not consistent with the 16S rDNA phylogenies [50, 51]. The isolates recovered from *Med. laciniata* were found to harbor an unusual *nodC* allele [45].

Table 3: Genes frequently used in MLSA of rhizobia

Gene	Gene Product	Rhizobial Population	References
16S rDNA	small subunit ribosomal RNA	*Bradyrhizobium* spp.	[42]
adeC3	adenine deaminase	*Ensifer medicae* (formerly *Sinorhizobium medicae*) *Ensifer meliloti* (formerly *Sinorhizobium meliloti*)	[41]
asd	aspartate-semialdehyde dehydrogenase	*Medicago*-nodulating rhizobia	[44, 45]

Table 3: cont….

atpD	subunit of ATP synthase	*Bradyrhizobium* spp.	[42, 46]
catC	catalase	*Ens. medicae* *Ens. meliloti*	[41]
dak	glycerone kinase	*Ens. medicae* *Ens. meliloti*	[41]
dgoA	galactonate dehydratase	*Ens. medicae* *Ens. meliloti*	[41]
dnaK	chaperone, a heat shock protein	*Bradyrhizobium* spp. *Ensifer* spp. *Microvirga lotononidis* *Microvirga lupini* *Microvirga zambiensis*	[40, 42, 52]
edd	phosphogluconate dehydratase	*Medicago*-nodulating rhizobia	[44, 45]
fixK	transcriptional regulator FixK	*Ens. medicae* *Ens. meliloti*	[41]
gabT	4-aminobutyrate amino-transferase	*Ens. medicae* *Ens. meliloti*	[41]
gap	glyceraldehyde-3-phosphate dehydrogenase	*Medicago*-nodulating rhizobia	[44, 45]
glnA	glutamine synthetase type I	*Ensifer* spp.	[40]
glnD	protein-PII uridylyl-transferase	*Medicago*-nodulating rhizobia	[44, 45]
glnII	glutamine synthetase II	*Bradyrhizobium* spp.	[42, 46]
gltA	citrate synthase	*Ensifer* spp.	[40]
gnd	6-phosphogluconate dehydrogenase	*Medicago*-nodulating rhizobia	[44, 45]
gyrB	β subunit of DNA gyrase	*Mic. lotononidis* *Mic. lupini* *Mic. zambiensis*	[52]
idh	*myo*-inositol dehydrogenase	*Ens. medicae* *Ens. meliloti*	[41]
ITS region	internal transcribed spacer of rRNA gene	*Bradyrhizobium* spp.	[42]
napB	nitrate reductase cytochrome *c*-type subunit	*Ens. medicae* *Ens. meliloti*	[41]
nifD	nitrogenase molybdenum-iron protein	*Ens. medicae* *Ens. meliloti*	[41]
nodC	*N*-acetylglucosaminyl-transferase	*Ens. medicae* *Ens. meliloti* *Medicago*-nodulating rhizobia	[41, 45]
nuoE1	NADH dehydrogenase I chain E protein	*Medicago*-nodulating rhizobia	[44, 45]

Table 3: cont….

ordL2	putative oxidoreductase protein	*Medicago*-nodulating rhizobia	[44, 45]
pdh	pyruvate dehydrogenase	*Ens. medicae* *Ens. meliloti*	[41]
recA	DNA strand exchange and recombination protein	*Bradyrhizobium* spp. *Mic. lotononidis* *Mic. lupini* *Mic. zambiensis* *Medicago*-nodulating rhizobia	[42, 44-46, 52]
rpoB	β subunit of DNA-dependent RNA polymerase	*Bradyrhizobium* spp. *Mic. lotononidis* *Mic. lupini* *Mic. zambiensis*	[46, 52]
SMa0198	ABC transporter, permease.	*Ens. medicae* *Ens. meliloti*	[41]
sucA	2-oxoglutarate dehydrogenase E1	*Medicago*-nodulating rhizobia	[44, 45]
thrC	threonine synthase	*Ensifer* spp.	[40]
zwf	glucose-6-phosphate 1-dehydrogenase	*Medicago*-nodulating rhizobia	[44, 45]

8.5. WHOLE GENOME SEQUENCE ANALYSIS OF RHIZOBIA

Whole genome sequence analysis provides the information needed to perform functional analysis of the genes as well as new insights into gene function, gene evolution and genome evolution by comparing the gene components and gene organization in the genomes among species [53]. The approach called "comparative genomics" is commonly used to encompass the capabilities of alignment and genome comparison. It has recently been used to analyze corresponding coding and non-coding regions from different species. Comparative analysis of a number of phylogenetically diverse genomes may provide clues about the selective pressures governing gene/operon clustering and may offer insights into mechanisms of evolution or show patterns in acquisition of foreign material *via* horizontal gene transfer [54]. Whole genome sequence can be achieved by the whole-genome shortgun method consisting of 4 steps: i) restriction enzymatic digestion or mechanical shearing of genomic DNA; ii) cloning of DNA fragments into the plasmid to construct a genome library; iii) purification of plasmid DNA; iv) sequencing of inserted DNA fragments. Alternatively, genomic DNA can be purified and PCR primers are designed to produce amplified PCR products representing regions in the whole genome. To obtain whole genome sequence data, PCR products are pooled and sequenced. Comparative genomics can be elucidated by using software packages for sequence assembly, quality assessment and gene identification. Whole genome sequences of rhizobia have been published for *Azorhizobium caulinodans* ORS571, *Bra. japonicum* USDA 110, *Bradyrhizobium* sp. BTAi1, *Bradyrhizobium* sp. ORS278, *Cupriavidus taiwanensis* LMG 19424, *Ensifer fredii* (formerly *Rhizobium* sp. and *Sinorhizobium fredii*) NGR234, *Ensifer meliloti* (formerly *Sinorhizobium meliloti*) 1021, *Mesorhizobium loti* MAFF303099, *Rhizobium etli* CFN42, *Rhi. etli* CIAT652 and *Rhizobium leguminosarum* biovar viciae 3841. Whole genome sequences of rhizobia consist of 1 or 2 large chromosomes ranging in sizes from 2.5 to 9.1 megabases (Mb) and, in some rhizobia, 1 to 6 megaplasmids ranging in sizes from 147.7 kb to 2.4 Mb. As revealed by comparative genomics, no gene was both common and specific to all rhizobia, thus suggesting that a unique shared genetic strategy does not support symbiosis of rhizobia with legumes. Instead, phylodistribution analysis of more than 200 *Ens. meliloti* known symbiotic genes indicated large and complex variations of their occurrence in rhizobia and non-rhizobia. This suggested first that rhizobia may use different strategies for symbiosis and second that some rhizobial symbiotic genes may be shared by non-rhizobia. However, some genes that were overrepresented in rhizobia could be selected. These include 96 genes of unknown function, 12 key symbiotic genes involved in nodulation (*nodA*, *nodC* and *nodD*) and nitrogen fixation (*nifD, nifK, nifE, nifN, nifB, fixA, fixC, fixX* and *fdxB*), 3 adenylate cyclase genes (*cyaH*, *cyaG1* and *cyaG2*), 2 genes encoding enzymes in production of phytohormone indole-3-acetic acid (IAA), a nitrate transporter gene (*nrtA*) and 2 glutathione-S-transferase (GST) genes [55]. The complete genome sequences of 2 symbiotic, photosynthetic *Bradyrhizobium* spp. BTAi1 and ORS278 lack *nod* genes that are specific for other rhizobia [56]. *C. taiwanensis* LMG19424 genome is significantly smaller than those of other available *Cupriavidus* genomes and also shows higher accumulation of insertion sequence (IS) than other *Cupriavidus*. These two features, *i.e.*, genome size reduction and IS accumulation, are expected for a plant symbiont that have recently passed an evolutionary bottleneck and adapted to a stable environment [55]. The IS accumulation presents as a common feature in rhizobia. The IS and transposon-like elements make up 18% of the symbiotic plasmid, pNGR234a, of *Ens. fredii* NGR234 [57]. A total of 167 genes for putative transposases were assigned to the genome of *Bra. japonicum* USDA 110. Of these, 133 were identified as IS [1]. The transposon elements account for 2.2% of the sequence of *Ens. meliloti* 1021 [58]. The replicons constituting whole genome of rhizobia are presented in Table **4**.

Table 4: The replicons constituting whole genome of rhizobia

Rhizobial Strain Replicon	Size (bp)*	Genbank Accession No.	General Feature	References
Azorhizobium caulinodans ORS571				
chromosome	5,369, 772	NC_009937; AP009384	i) has an average guanine-cytosine (GC) content of 67.0%. ii) 76.1 % are functional genes, 20.2% are hypothetical genes and 3.7% are unique. iii) contains regulatory/transporter genes and genes possibly involved in contacting a host.	[59]
total	5,369,772			
Bradyrhizobium japonicum USDA 110				
chromosome	9,105,828	NC_004463; BA000040	i) has an average GC content of 64.1%. ii) 52.3 % are functional genes, 30.1% are hypothetical genes and 17.6% are unique.	[53]
total	9,105,828			
Bradyrhizobium sp. BTAi1				
chromosome	8,264,687	NC_009485; CP000494	lacks nodulation (*nod*) genes that are specific for other rhizobia.	[56]
plasmid pBBta01	228,826	NC_009475; CP000495	lacks *nod* genes that are specific for other rhizobia.	[56]
total	8,493,513			
Bradyrhizobium sp. ORS278				
chromosome	7,456,587	NC_009445	lacks *nod* genes that are specific for other rhizobia.	[56]
total	7,456,587			
Cupriavidus taiwanensis LMG 19424				
chromosome 1	3,416,911	CU633749	i) has an average GC content of 67.5%. ii) contains genes for secretion systems and interaction with a host.	[55]

Table 4: cont...

chromosome 2	2,502,411	CU633750	i) has an average GC content of 67.9%. ii) contains genes encoding ATP sulfurylase, APS kinase and adenyl sulfate kinase.	[55]
plasmid pRalta	557,200	CU633751	i) has an average GC content of 59.6%. ii) is symbiotic plasmid (pSym) that contains most of the genes required for symbiosis.	[55]
total	6,476,522			
Ensifer fredii (formerly Rhizobium sp. and Sinorhizobium fredii) NGR234				
chromosome	3,925,702	NC_012587; CP001389	i) has an average GC content of 62.8%. ii) contains 5 nitrogen fixation (*nif*) and *nod* genes. iii) contains genes for cellular growth.	[60]
plasmid pNGR234a	536,165	U00090	i) has an average GC content of 58.4%. ii) is pSym.	[60]
plasmid pNGR234b	2,430,033	NC_000914	i) has an average GC content of 62.3%. ii) contains 8 *nif* and *nod* genes. ii) contains genes for a few essential functions and exopolysaccharide synthesis.	[57, 60]
total	6,891,900			
Ensifer meliloti (formerly Sinorhizobium meliloti) 1021				
chromosome	3,653,135	NC_003047	i) has an average GC content of 62.7%. ii) 59.0 % are functional genes, 29.0% are hypothetical genes and 12.0% are unique. iii) contains house keeping genes and genes for mobility, chemotaxis process, plant interaction and stress response.	[58]
plasmid pSymA	1,354,226	NC_003037; AE006469	i) has an average GC content of 58.0% ii) is pSym. iii) contains genes likely to be involved in nitrogen/ carbon metabolism, transport, stress, resistance responses and adaptation to environment.	[61]

Table 4: cont….

plasmid pSymB	1,683,333	NC_003078; AL591985	i) has an average GC content of 62.4%. ii) is pSym. iii) contains 2 essential genes and genes for solute uptake systems and polysaccharide synthesis.	[62]
total	6,691,694			
Mesorhizobium loti **MAFF303099**				
chromosome	7,036,071	NC_002678; BA000012	i) 54.0 % are functional genes, 21.0% are hypothetical genes and 25.0% are unique. ii) contains 611-kb symbiosis island.	[63]
plasmid pMLa	351,911	NC_002679; BA000013	i) contains genes for ABC-transporter system, phosphate assimilation, two-component system, DNA replication and conjugation. ii) contains only one gene for nodulation.	[63]
plasmid pMLb	208,315	NC_002682; AP003017	contains genes for ABC-transporter system, phosphate assimilation, two-component system, DNA replication and conjugation.	[63]
total	7,596,297			
Rhizobium etli **CFN42**				
chromosome	4,381,608	NC_007761; CP000133	i) has an average GC content of 61.3%. ii) 71.1% are functional genes, 23.7% are hypothetical genes and 5.2% are unique. iii) contains several genes for symbiosis.	[64]
plasmid p42a	194,229	NC_007762; CP000134	i) has an average GC content of 58.0% ii) 68.7% are functional genes, 24.2% are hypothetical genes and 7.1% are unique.	[64]
plasmid p42b	184,338	NC_007763; CP000135	i) has an average GC content of 61.8%. ii) 83.6% are functional genes, 14.0% are hypothetical genes and 2.4% are unique. iii) contains lipopolysaccharide (*lps*) gene involved in symbiosis.	[64]

Table 4: cont...

plasmid p42c	250,948	NC_007764; CP000136	i) has an average GC content of 61.5%. ii) 76.1% are functional genes, 19.2% are hypothetical genes and 4.7% are unique.	[64]
plasmid p42d	371,225	NC_004041; U80928	i) has an average GC content of 58.4%. ii) 67.0% are functional genes, 16.9% are hypothetical genes and 16.1% are unique. iii) is pSym.	[65]
plasmid p42e	505,334	NC_007765; CP000137	i) has an average GC content of 61.7%. ii) 69.1% are functional genes, 25.7% are hypothetical genes and 5.2% are unique.	[64]
plasmid p42f	642,517	NC_007766; CP000138	i) has an average GC content of 61.2%. ii) 70.3% are functional genes, 23.4% are hypothetical genes and 6.3% are unique. iii) contains nitrogen fixation (*fix*) genes.	[64]
total	6,530,229			
Rhizobium etli **CIAT652**				
chromosome	4,513,324	NC_010994; CP001074		[66]
plasmid pCIAT652a	414,090	NC_010998; CP001075	contains genes in common with p42e of *Rhi. etli* CFN42.	[66]
plasmid pCIAT652b	429,111	NC_010996; CP001076	i) is pSym. ii) contains genes in common with p42d of *Rhi. etli* CFN42.	[66]
plasmid pCIAT652c	1,091,523	NC_010997; CP001077	contains genes in common with p42b and p42f of *Rhi. etli* CFN42.	[66]
total	6,448,048			
Rhizobium leguminosarum **biovar viciae 3841**				
chromosome	5,057,142	NC_008380; AM236080	has an average GC content of 61.1%.	[67]
plasmid pRL7	151,564	NC_008382; AM236081	i) has an average GC content of 57.6%. ii) is transmissible plasmid.	[67]
plasmid pRL8	147,463	NC_008383; AM236082	i) has an average GC content of 58.7%. ii) is transmissible plasmid.	[67]

Table 4: cont...

plasmid pRL9	352,782	NC_008379; AM236083	has an average GC content of 61.0%.	[67]
plasmid pRL10	488,135	NC_008381; AM236084	i) has an average GC content of 59.6%. ii) is pSym	[67]
plasmid pRL11	684,202	NC_008384; AM236085	i) has an average GC content of 61.0%. ii) contains genes similar to those of the chromosome.	[67]
plasmid pRL12	870,021	NC_008378; AM236086	i) has an average GC content of 61.0%. ii) contains genes similar to those of the chromosome.	[67]
total	7,751,309			

bp: base pairs.

REFERENCES

[1]   Clarridge III JE. Impact of 16S rRNA gene sequence analysis for identification of bacteria on clinical microbiology and infectious diseases. Clin Microbiol Rev 2004; 17: 840-62.

[2]   van Berkum P, Eardly BD. Molecular Evolutionary Systematics of the Rhizobiaceae. In: Spaink HP, Kondorosi A, Hooykaas PJJ, Eds. The Rhizobiaceae: Molecular Biology of Model Plant-Associated Bacteria. Dordrecht, Kluwer Academic Publishers, 1998; pp. 1-24.

[3]   Young JPW, Downer HL, Eardly BD. Phylogeny of the phototropic *Rhizobium* strain BTAi1 by polymerase chain reaction-based sequencing of a 16S rRNA gene segment. J Bacteriol 1991; 173: 2271-77.

[4]   Laguerre G, Mavingui P, Allard MR, *et al.* Typing of rhizobia by PCR DNA fingerprinting and PCR-restriction fragment length polymorphism analysis of chromosomal and symbiotic gene regions: application to *Rhizobium leguminosarum* and its different biovars. Appl Envir Microbiol 1996; 62: 2029-36.

[5]   Saitou N, Nei M. The neighbor-joining method: A new method for reconstructing phylogenetic trees. Mol Biol Evol 1987; 4: 406-25.

[6]   Higgins DG, Bleasby AJ, Fuchs R. CLUSTAL V: improved software for multiple sequence alignment. Comput Appl Biosci 1992; 8: 189-91.

[7]   Thompson JD, Gibson TJ, Plewniak F, *et al.* The CLUSTALX windows interface: flexible strategies for multiple sequence alignment aided by quality analysis tools. Nuclcic Acids Res 1997; 25: 4876-82.

[8]   Tamura K, Nei M, Kumar S. Prospects for inferring very large phylogenies by using the neighbor-joining method. Proc Natl Acad Sci USA. 2004; 101: 11030-35.

[9]   Kumar S, Tamura K, Nei M. MEGA 3: integrated software for molecular evolutionary genetics analysis and sequence alignment. Brief Bioinform 2004; 5: 150-63.

[10] Tamura K, Dudley J, Nei M, Kumar S. MEGA 4: molecular evolutionary genetics analysis (MEGA) software version 4.0. Mol Biol Evol 2007; 24: 1596-99.

[11]   Young JPW, Haukka K. Diversity and phylogeny of rhizobia. New Phytol 1996; 133: 87-94.

[12]   Barrett CF, Parker MA. Coexistence of *Burkholderia*, *Cupriavidus*, and *Rhizobium* sp. nodule bacteria on two *Mimosa* spp. in Costa Rica. Appl Envir Microbiol 2006; 72: 1198-06.

[13]   Chen WM, Laevens S, Lee TM, *et al. Ralstonia taiwanensis* sp. nov., isolated from root nodules of *Mimosa* species and sputum of a cystic fibrosis patient. Int J Syst Evol Microbiol 2001; 51: 1729-35.

[14]   Chen WM, de Faria SM, Straliotto R, *et al.* Proof that *Burkholderia* strains form effective symbioses with legumes: a study of novel *Mimosa*-nodulating strains from South America. Appl Envir Microbiol 2005; 71: 7461-71.

[15]   Chen WM, James EK, Coenye T, *et al. Burkholderia mimosarum* sp. nov., isolated from root nodules of *Mimosa* spp. from Taiwan and South America. Int J Syst Evol Microbiol 2006; 56: 1847-51.

[16]   Ngom A, Nakagawa Y, Sawada H, *et al.* A novel symbiotic nitrogen-fixing member of the *Ochrobactrum* clade isolated from root nodules of *Acacia mangium*. J Gen Appl Microbiol 2004; 50: 17-27.

[17]   Rivas R, Valazquez E, Willems A, *et al.* A new species of *Devosia* that forms a unique nitrogen-fixing root-nodule symbiosis with the aquatic legume *Neptunia natans* (L.f.) Druce. Appl Envir Microbiol 2002; 68: 5217-22.

[18]   Sy A, Giraud E, Jourand P, *et al.* Methylotrophic *Methylobacterium* bacteria nodulate and fix nitrogen in symbiosis with legumes. J Bacteriol 2001; 183: 214-20.

[19]   Valverde A, Velazquez E, Gutierrez C, *et al. Herbaspirillum lusitanum* sp. nov., a novel nitrogen-fixing bacterium associated with root nodules of *Phaseolus vulgaris*. Int J Syst Evol Microbiol 2003; 53: 1979-83.

[20]   Valverde A, Velazquez E, Fernandez-Santos F, *et al. Phyllobacterium trifolii* sp. nov., nodulating *Trifolium* and *Lupinus* in Spanish soils. Int J Syst Evol Microbiol 2005; 55: 1985-89.

[21]   van Berkum P, Eardly BD. The aquatic budding bacterium *Blastobacter denitrificans* is a nitrogen-fixing symbionts of *Aeschynomene indica*. Appl Envir Microbiol 2002; 68: 1132-36.

[22]   Fuentes JB, Abe M, Uchiumi T, *et al.* Symbiotic root nodule bacteria isolated from yam bean (*Pachyrhizus erosus*). J Gen Appl Microbiol 2002; 48: 181-91.

[23]   Sawada H, Ieki H, Oyaizu H, Matsumoto S. Proposal for rejection of *Agrobacterium tumefaciens* and revised descriptions for the genus *Agrobacterium* and for *Agrobacterium radiobacter* and *Agrobacterium rhizogenes*. Int J Syst Bacteriol 1993; 43: 694-02.

[24]   Young JM, Kuykendall LD, Martinez-Romero E, *et al.* A revision of *Rhizobium* Frank 1889, with an emended description of the genus, and the inclusion of all species of *Agrobacterium* Conn 1942 and *Allorhizobium undicola* de Lajudie *et al.* 1998 as new combinations: *Rhizobium radiobacter, R. rhizogenes, R. rubi, R. undicola* and *R. vitis*. Int J Syst Evol Microbiol 2001; 51: 89-03.

[25]   Saito A, Mitsui H, Hattori R, *et al.* Slow-growing and oligotrophic soil bacteria phylogenetically close to *Bradyrhizobium japonicum*. FEMS Microbiol Ecol 1998; 25: 277-86.

[26]   Sawada H, Kuykendall LD, Young JM. Changing concepts in the systematics of bacterial nitrogen-fixing legume symbionts. J Gen Appl Microbiol 2003; 49: 155-79.

[27]   Willems A, Coopman R, Gillis M. Phylogenetic and DNA-DNA hybridization analyses of *Bradyrhizobium* species. Int J Syst Evol Microbiol 2001; 51: 111-17.

[28] Shamseldin A, El-Saadani M, Sadowsky MJ, An CS. Rapid identification and discrimination among Egyptian genotypes of *Rhizobium leguminosarum* bv. *viciae* and *Sinorhizobium meliloti* nodulating faba bean (*Vicia faba* L.) by analysis of *nodC*, ARDRA, and rDNA sequence analysis. Soil Biol Biochem 2009; 41: 45-53.

[29] Pongsilp N, Teaumroong N, Nuntagij A, *et al.* 2002. Genetic structure of indigenous non-nodulating and nodulating populations of *Bradyrhizobium* in soils from Thailand. Symbiosis 33: 39-58.

[30] Rashid MH, Sattar MA, Uddin MI, Young JPW. Molecular characterization of symbiotic root nodulating rhizobia isolated from lentil (*Lens culinaris* Medik.). EJEAFChe 2009; 8: 602-12.

[31] Silva C, Vinuesa P, Eguiarte LE, *et al. Rhizobium etli* and *Rhizobium gallicum* nodulate common bean (*Phaseolus vulgaris*) in a traditionally managed Milpa Plot in Mexico: population genetics and biogeographic implications. Appl Envir Microbiol 2003; 69: 884-93.

[32] Nimnoi P, Lumyong S, Pongsilp N. Impact of rhizobial inoculants on rhizosphere bacterial communities of three medicinal legumes assessed by denaturing gradient gel electrophoresis (DGGE). Ann Microbiol 2011; 61: 237-45.

[33] Ramsubhag A, Umaharan P, Donawa A. Partial 16S rRNA gene sequence diversity and numerical taxonomy of slow growing pigeon pea (*Cajanus cajan* L. Millsp) nodulating rhizobia. FEMS Microbiol Lett 2002; 216: 139-44.

[34] Terefework Z, Nick G, Suomalainen S, *et al.* Phylogeny of *Rhizobium galegae* with respect to other rhizobia and agrobacteria. Int J Syst Bacteriol 1998; 48: 349-56.

[35] Bromfield ESP, Tambong JT, Cloutier S, *et al. Ensifer*, *Phyllobacterium* and *Rhizobium* species occupy nodules of *Medicago sativa* (alfalfa) and *Melilotus alba* (sweet clover) grown at a Canadian site without a history of cultivation. Microbiol 2010; 156: 505-20.

[36] Berridge BR, Fuller JD, de Azavedo J, *et al.* Development of specific nested oligonucleotide PCR primers for the *Streptococcus iniae* 16S-23S ribosomal DNA intergenic spacer. J Clin Microbiol 1998; 36: 2778-81.

[37] Jensen MA, Webster JA, Straus N. Rapid identification of bacteria on the basis of polymerase chain reaction-amplified ribosomal DNA spacer polymorphisms. Appl Envir Microbiol 1993; 59: 945-52.

[38] de Oliveira VM, Coutinho HLC, Sobral BWS, *et al.* Discrimination of *Rhizobium tropici* and *R. leguminosarum* strains by PCR-specific amplification of 16S-23S rDNA spacer region fragments and denaturing gradient gel electrophoresis (DGGE). Lett Appl Microbiol 1999; 28: 137-41.

[39] Tan Z, Hurek T, Vinuesa P, *et al.* Specific detection of *Bradyrhizobium* and *Rhizobium* strains colonizing rice (*Oryza sativa*) roots by 16S-23S ribosomal DNA intergenic spacer-targeted PCR. Appl Envir Microbiol 2001; 67: 3655-64.

[40] Martens M, Delaere M, Coopman R, *et al.* Multilocus sequence analysis of *Ensifer* and related taxa. Int J Syst Evol Microbiol 2007; 57: 489-03.

[41] van Berkum P, Elia P, Eardly BD. Application of multilocus sequence typing to study the genetic structure of megaplasmids in *Medicago*-nodulating rhizobia. Appl Envir Microbiol 2010; 76: 3967-77.

[42] Menna P, Barcellos FG, Hungria M. Phylogeny and taxonomy of a diverse collection of *Bradyrhizobium* strains based on multilocus sequence analysis of the 16S rRNA gene, ITS region and *glnII*, *recA*, *atpD* and *dnaK* genes. Int J Syst Evol Microbiol 2009; 59: 2934-50.

[43] Stepkowski T, Czaplinska M, Miedzinska K, Moulin L. The variable part of the *dnaK* gene as an alternative marker for phylogenetic studies of rhizobia and related alpha Proteobacteria. Syst Appl Microbiol 2003; 26: 483-94.

[44] van Berkum P, Elia P, Eardly BD. Multilocus sequence typing as an approach for population analysis of *Medicago*-nodulating rhizobia. J Bacteriol 2006; 188: 5570-77.

[45] van Berkum P, Badri Y, Elia P, *et al.* Chromosomal and symbiotic relationships of rhizobia nodulating *Medicago truncatula* and *M. laciniata.* Appl Envir Microbiol 2007; 73: 7597-04.

[46] Vinuesa P, Rojas-Jimenez K, Contreras-Moreira B, *et al.* Multilocus sequence analysis for asssessment of the biogeography and evolutionary genetics of four *Bradyrhizobium* species that nodulate soybeans on the Asiatic Continent. Appl Envir Microbiol 2008; 74: 6987-96.

[47] Georgiadis MM, Komiya H, Chakrabarti P, *et al.* Crystallographic structure of the nitrogenase iron protein from *Azotobacter vinelandii.* Science 1992; 257: 1653-59.

[48] Geremia RA, Mergaert P, Geelen D, *et al.* The NodC protein of *Azorhizobium caulinodans* is an N-acetylglucosaminyltransferase. Proc Natl Acad Sci USA. 1994; 91: 2669-73.

[49] Pongsilp N, Leelahawonge C. 2010. Root-nodule symbionts of *Derris elliptica* Benth. are members of three distinct genera *Rhizobium*, *Sinorhizobium* and *Bradyrhizobium*. Int J Integr Biol 9: 37-42.

[50] Laguerre G, Nour SM, Macheret V, *et al.* Classification of rhizobia based on *nodC* and *nifH* gene analysis reveals a close phylogenetic relationship among *Phaseolus vulgaris* symbionts. Microbiol 2001; 147: 981-93.

[51] Vinuesa P, Leon-Barrios M, Silva C, *et al. Bradyrhizobium canariense* sp. nov., an acid-tolerant endosymbiont that nodulates endemic genistoid legumes (Papilionoideae: Genisteae) from the Canary Islands, along with *Bradyrhizobium japonicum* bv. *genistearum*, *Bradyrhizobium* genospecies alpha and *Bradyrhizobium* genospecies beta. Int J Syst Evol Microbiol 2005; 55: 569-75.

[52] Ardley JK, Parker MA, de Meyer SE, *et al. Microvirga lupini* sp. nov., *Microvirga lotononidis* sp. nov., and *Microvirga zambiensis* sp. nov. are alphaproteobacterial root nodule bacteria that specifically nodulate and fix nitrogen with geographically and taxonomically separate legume hosts. Int J Syst Evol Microbiol 2011; doi: 10.1099/ijs.0.035097-0.

[53] Kaneko T, Nakamura Y, Sato S, *et al.* Complete genomic sequence of nitrogen-fixing symbiotic bacterium *Bradyrhizobium japonicum* USDA 110. DNA Res 2002; 9: 189-97.

[54] Chain P, Kurtz S, Ohlebusch E, Slezak T. An applications-focused review of comparative genomics tools: capabilities, limitations and future challenges. Brief Bioinform 2003; 4: 105-23.

[55] Amadou C, Pascal G, Mangenot S, *et al.* Genome sequence of the β-rhizobium *Cupriavidus taiwanensis* and comparative genomics of rhizobia. Genome Res 2008; 18: 1472-83.

[56] Giraud E, Moulin L, Vallenet D, *et al.* Legumes symbioses: absence of Nod genes in photosynthetic bradyrhizobia. Science 2007; 316: 1307-12.

[57] Freiberg C, Fellay R, Bairoch A, *et al.* Molecular basis of symbiosis between *Rhizobium* and legumes. Nature 1997; 387: 394-01.

[58] Capela D, Barloy-Hubler F, Gouzy J, *et al.* Analysis of the chromosome sequence of the legume symbiont *Sinorhizobium meliloti* strain 1021. Proc Natl Acad Sci USA. 2001; 98: 9877-82.

[59]   Lee KB, de Backer P, Aono T, *et al.* The genome of the versatile nitrogen fixer *Azorhizobium caulinodans* ORS571. BMC Genomics 2008; 9: 271.

[60]   Schmeisser C, Liesegang H, Krysciak D, *et al. Rhizobium* sp. strain NGR234 possesses a remarkable number of secretion systems. Appl Envir Microbiol 2009; 75: 4035-45.

[61]   Barnett MJ, Fisher RF, Jones T, *et al.* Nucleotide sequence and predicted functions of the entire *Sinorhizobium meliloti* pSymA megaplasmid. Proc Natl Acad Sci USA. 2001; 98: 9883-88.

[62]   Finan TM, Weidner S, Wong K, *et al.* The complete sequence of the 1,683-kb pSymB megaplasmid from the N_2-fixing endosymbiont *Sinorhizobium meliloti*. Proc Natl Acad Sci USA. 2001; 98: 9889-94.

[63]   Kaneko T, NakamuraY, Sato S, *et al.* Complete genome structure of the nitrogen-fixing symbiotic bacterium *Mesorhizobium loti*. DNA Res 2000; 7: 331-38.

[64]   Gonzalez V, Santamaria RI, Bustos P, *et al.* The partitioned *Rhizobium etli* genome: genetic and metabolic redundancy in seven interacting replicons. Proc Natl Acad Sci USA. 2006; 103: 3834-39.

[65]   Gonzalez V, Bustos P, Ramirez-Romero MA, *et al.* The mosaic structure of the symbiotic plasmid of *Rhizobium etli* CFN42 and its relation to other symbiotic genome compartments. Genome Biol 2003; 4: R36.

[66]   Gonzalez V, Acosta JL, Santamaria RI, *et al.* Conserved symbiotic plasmid DNA sequences in the multireplicon pangenomic structure of *Rhizobium etli*. Appl Envir Microbiol 2010; 76: 1604-14.

[67]   Young JP, Crossman LC, Johnston AW, *et al.* The genome of *Rhizobium leguminosarum* has recognizable core and accessory components. Genome Biol 2006; 7: R34.

CHAPTER 9

Application of Rhizobia in Agriculture

Neelawan Pongsilp[*]

Department of Microbiology, Faculty of Science, Silpakorn University, Nakhon Pathom, Thailand

Abstract: Rhizobia are of enormous agricultural and economic values because they provide the major source of nitrogen input in agricultural soils. Beside nitrogen fixation, many rhizobial strains exert plant-growth-promoting traits such as the production of phytohormones, siderophores and 1-aminocyclopropane-1-carboxylic acid (ACC) deaminase as well as the solubilization of inorganic phosphate. These make rhizobia become valuable for both legumes and non-legumes. Effective rhizobial strains have been screened and used as inoculants for improving plant growth. The application of rhizobia as biofertilizer ensures success in crop productivity and reduces the need for artificial fertilizers that are expensive and cause environmental problems. Co-inoculations of rhizobia and other plant-growth-promoting rhizobacteria (PGPR) have shown the significant increase in plant-growth promotion. Several rhizobial strains have also been reported for their bioremediation abilities.

Keywords: Acetylene Reduction Activity (ARA), Biofertilizer, Biological nitrogen fixation (BNF), Bioremediation, Co-inoculation, Endophyte, Inoculant, Plant-growth-promoting rhizobacteria (PGPR), Rhizobia, Symbiotic effectiveness.

9.1. RHIZOBIA USED AS BIOFERTILIZER

Rhizobia attract great interest by providing the nitrogen (N) requirements of their host plants and reducing the need for artificial fertilizers which are expensive and cause environmental problems. An examination of the history of biological nitrogen fixation (BNF) shows that the interest generally has focused on the symbiotic system of leguminous plants and rhizobia, because these associations have the greatest quantitative impact on the N cycle [1]. Besides N fixation, thcy can also promote plant growth by mechanisms such as the production of plant growth regulators, siderophores, ammonia and some enzymes as well as the ability to solubilize inorganic phosphate. For agronomic utility, inoculation of

*Address correspondence to Neelawan Pongsilp: Department of Microbiology, Faculty of Science, Silpakorn University, Nakhon Pathom, Thailand; Tel: +66-34-245337; Fax: +66-34-245336-37; E-mail: neelawan@su.ac.th

plants with specific rhizobia at a much higher concentration than those normally found in soil is necessary to take advantages of their beneficial properties for plant yield enhancement. Rhizobia have been successfully used as promising biofertilizer for agricultural improvement. In Iran soils with N deficiency, *Rhizobium* could increase yield of *Phaseolus vulgaris* (common bean) at a low cost and preserve water resources from pollution caused by chemical nitrate fertilizer [2]. Inoculation with rhizobial strain is a promising fertilizer because it is cheap, easy to handle and improves plant growth and seed quality.

It is essential to establish the rhizobia-legume symbioses in the production of legume crops. Inoculation of rhizobia has successfully been used for improving growth and yield of legume crops world wide. *Bradyrhizobium japonicum* USDA 110 has been used extensively as the inoculant strain for *Glycine max* (soybean) because of its superior symbiotic N fixation and its tolerance to stress conditions [3]. The strain USDA 110 improved soybean growth with plant dry weight and total N content up to 1.36 and 3.51 times, respectively, as compared to the uninoculated controls. *Ensifer meliloti* (formerly *Rhizobium meliloti*) 042B has been reported for the equal symbiotic effectiveness to USDA 110 [4]. The 10 strains of rhizobia nodulating *Vigna radiata* (mungbean) conferred N fixation efficiency in terms of acetylene reduction activity (ARA) 232.00 to 717.54 times over the uninoculated controls. The most effective strains enhanced plant dry weight up to 1.98 times over the uninoculated controls [5]. *Rhizobium* inoculations increased plant height, first pod height, number of branches, pod number per plant, seed number per plant and grain yield of *Cicer arietinum* L. (chickpea) [6]. Inoculations with *Rhizobium leguminosarum* strains had a significant effect on *Pisum sativum* (pea) in terms of plant height, number of branches, root and shoot dry weight, number of nodule, seed yield, biomass yield, number of pod, crude protein rate and phosphorus (P) content of seed. The positive relationship of yield and growth parameters with inoculation could be related to the N fixation ability of nodules, which consequently resulted in increased growth and yield [7]. In both non-saline and saline (50 and 100 mM NaCl) conditions, inoculations with *Rhi. leguminosarum* biovar ciceri strains isolated from *Cic. arietinum* L. significantly increased root and shoot dry weights, root-to-shoot ratio (RSR), nodule number, nodule dry weight, chlorophyll content,

N content of the plant and the amount of fixed N, equal to or higher than standard culture and N application [8]. *Rhizobium* inoculation supplemented with L-tryptophan (10^{-4} M) gave the most promising results and significantly increased the growth of *V. radiata* with the plant height, number of nodules per plant, nodular mass per plant, number of grains per pod, number of pods per plant, total plant biomass, grain yield and 1,000-grain weight up to 28%, 80%, 77%, 46%, 54%, 58%, 57% and 17%, respectively, as compared to the uninoculated controls. Similarly, N concentration in grains was also increased significantly at this L-tryptophan level by the rhizobial inoculation [9]. The application of *Rhizobium* inoculation significantly affected the plant height, root length, number of pods per plant, number of seeds per plant, leaf protein, seed protein and plant yield of *Dolichos biflorus* (horsegram) [10]. The 20 strains of rhizobia nodulating *Indigofera tinctoria* (true indigo) enhanced plant dry weight and ARA up to 1.03 to 10.55 and 5.82 to 10.91 times, respectively, over the uninoculated controls (Neelawan Pongsilp and Achara Nuntagij, unpublished data).

Rhizobia have also been found to be endophytes of non-legumes such as barley, canola, corn (maize), rice, wheat and other cereals without forming any nodule-like structure or causing any disease symptoms [11-16]. *Bradyrizobium japonicum* Soy 213 conferred the highest stimulatory effect (60%) on dry matter yield of *Raphanus sativus* (radish) as compared to the uninoculated controls. On average, about 25% of all *Rhizobium* and *Bradyrhizobium* strains tested stimulated the growth of *Rap. sativus* (20% or more increase) [17]. *Sinorhizobium* sp. AS014 stimulated the highest yield of *Rap. sativus* and *Brassica oleracea*, with averages up to 5.33-fold and 5.63-fold over the uninoculated controls for root yield of *Rap. sativus* and *Brassica oleracea*, respectively, also increased up to 3.34-fold and 2.95-fold over the uninoculated controls for shoot yield of *Rap. sativus* and *Brassica oleracea*, respectively [18]. Over 60 strains of *Rhizobium etli* biovar phaseoli were found to be natural endophytes of *Zea mays* (maize, corn). Increases in plant dry weight upon *Rhi. etli* inoculation were recorded with some of the maize land races, and these increases may be related to plant-growth-promotion effects [19]. *Sinorhizobium* spp. and *Mesorhizobium* spp. with phosphate-solubilization activity increased dry weight and P content of *Z. mays* fertilized with insoluble P [20]. The dynamic infection process of healthy *Oryza sativa* (rice) plant tissues by 4 species of rhizobia, including *Azorhizobium caulinodans*, *Ens. meliloti*, *Mesorhizobium haukuii* and *Rhi. leguminosarum* began with surface colonization of the rhizoplane (especially at lateral root emergence), followed by endophytic colonization within roots, and then ascended endophytic migration into the stem base, leaf sheath, and leaves where they developed high populations [21]. The 266 strains of *Rhizobium* and *Bradyrhizobium* were examined for plant-growth-promoting traits. It was found that the strains produced indole-3-acetic acid (IAA), siderophores and cyanide (HCN), solubilized phosphate as well as exhibited antagonistic effects toward many plant pathogenic fungi. Inoculation with *Bradyrhizobium* sp., *Rhi. leguminosarum* biovar trifolii and *Rhizobium* sp. increased rice grain and straw yield by 8% to 22% and 4% to 19%, respectively, at different N rates. It was also found that rhizobial inoculation increased N, P and potassium (K) uptake by 10% to 28%. Inoculation also increased iron (Fe) uptake in rice by 15% to 64% [22]. The 4 strains of *Mesorhizobium ciceri*, 4 strains of *Rhi. leguminosarum* and 3 strains of *Rhizobium phaseoli* were tested for their potential to promote growth and yield of rice. All the tested rhizobial strains have the potential to enhance the growth and yield of rice. *Rhi. leguminosarum* LSI-29 caused maximum increases in cases of number of tiller (46%), paddy yield (43%), plant biomass (18%), straw dry weight (45%) and 1,000-grain (25%), as compared to the uninoculated controls. *Rhi. phaseoli* A2 caused maximum increases in cases of plant height (28%) and number of grains panicle (29%), as compared to the uninoculated controls. Both strains gave significant increases in N, P and K contents of paddy [23]. *Azorhizobium caulinodans* and *Rhi. leguminosarum* biovar trifolii were found to promote the growth of rice plants [15, 24]. The photosynthetic *Bradyrhizobium* strains were found to be natural true endophytes of *Oryza breviligulata* (African wild rice). An endophytic strain increased the shoot growth and grain yield of *O. breviligulata* by 20%. The photosynthetic *Bradyrhizobium* sp. ORS278 extensively colonized the root surface, followed by intercellular and rarely intracellular, bacterial invasion of the rice roots [25].

9.2. CO-INOCULATIONS OF RHIZOBIA AND OTHER PLANT-GROWTH-PROMOTING RHIZOBACTERIA (PGPR)

Enhancement of legume N fixation by co-inoculations of rhizobia with some plant-growth-promoting rhizobacteria (PGPR) is a way to improve nitrogen availability in sustainable agriculture production systems. There are many reports depicting plant-growth promotion when rhizobia are co-inoculated with some other PGPR. The use of mixed biofertilizers is advocated to get the maximum benefits due to additive and synergistic effect [26]. Co-inoculation of *Rhizobium* sp. and arbuscular mycorrhizal (AM) fungus, *Glomus mosseae*, showed a limited improvement on *Stylosanthez hamata*. However, the improvement of growth and N-fixing ability of *Sty. hamata* did not only depend on the inoculation but also required the application of P source [27]. Combined inoculation of *Rhizobium*, a phosphate solubilizing *Bacillus megaterium* and a biocontrol fungus *Trichoderma* sp., showed increased germination, nutrient uptake, plant height, number of branches, nodulation, plant yield and total biomass of *Cic. arietinum* L. under glasshouse and field conditions, as compared to either individual inoculations or uninoculated controls [28]. Under field conditions, *Cic. arietinum* L. inoculated with *Mes. ciceri* C-2/2, in single or dual inoculations with a phosphate-solubilizing bacterium, *Pseudomonas jessenii* PS06, produced the maximum increases in nodule fresh weight, nodule number and shoot N content. The result suggests that *Mes. ciceri* C-2/2 can act synergistically with *Pse. jessenii* PS06 in promoting chickpea growth [29]. Co-inoculations of *Rhizobium* and a sulfur-oxidizing bacterium, tentatively identified as *Thiobacillus*, yielded the promising results on *Arachis hypogaea* L. (groundnut) with increasing plant biomass, nodule number, nodule dry weight and pod yield [30]. Under field conditions, *Pseudomonas fluorescens* and *Azospirillum lipoferum* strains positively affected symbiotic performance of *Rhi. phaseoli*. Co-inoculations of *Pha. vulgaris* with *Rhi. phaseoli* Rb-133 and *Pse. fluoresescens* P93 resulted in higher number of seed per pod, weight of seed per plant, protein seed yield and seed yield [2]. Co-inoculations of *Rhizobium* sp. and each of *Bacillus* strains showed plant-growth promotion on *Cajanus cajan* (pigeon pea) with respect to the increases in plant fresh weight, chlorophyll content, nodule number and nodule fresh weight. Under iron starved conditions, a strain of *Rhizobium* enhanced plant growth in the presence of *Bacillus* strains capable of producing siderophores, indicating that siderophore mediated interactions may be underlying mechanism of beneficial effect of *Bacillus* strains on nodulation by *Rhizobium* sp. [31]. In the controlled environment and in the field trials, single, dual and triple inoculations of *Cic. arietinum* L. with *Rhizobium*, N-fixing *Bacillus subtilis* and phosphate-solubilizing *Bac. megaterium* significantly increased plant height, shoot, root and nodule dry weight, %N, chlorophyll content, pod number, seed yield, total biomass yield, and seed protein content, as compared with the uninoculated controls, equal to or higher than the treatments supplied with N, P and N together with P. In the fields, all the combined treatments containing *Rhizobium* were better for nodulation than the use of *Rhizobium* alone [32]. Co-inoculation of *Rhizobium* and a free-living nitrogen fixing bacterium, *Azospirillum*, enhanced nodule number as compared to a single *Rhizobium* inoculation [33]. The effect of *Azospirillum* on the *Rhizobium*-legume symbiosis was found to be host genotype-dependent. Co-inoculation of *Rhizobium* and *Azospirillum* increased the amount of fixed N and the yield of *Pha. vulgaris* genotype DOR364 across all sites as compared to a single *Rhizobium* inoculation. On the contrary, a negative effect of this co-inoculation on yield and N fixation was observed for *Pha. vulgaris* genotype BAT477 on most of the sites as compared to a single *Rhizobium* inoculation [34]. Co-inoculation of *Rhizobium* sp. and an AM fungus, *Glomus deserticola*, exerted a synergic effect on *Robinia pseudoacacia* L. (black locust) because the co-inoculation produced significantly higher shoot biomass, nodule biomass and N fixation, as measured by ARA, (at least 93%, 50%, and 192%, respectively) than any single inoculation treatment and than the uninoculated controls, which were nodulated by soil-born rhizobia. Mycorrhizal colonization did not improve the growth of control plants, but it significantly improved the specific nitrogenase activity of plants inoculated also with *Rhizobium* sp. by 77%. It is concluded that prenodulation of plants is necessary for better growth on harsh substrates and mycorrhizal colonization helps nodulated plants reach their P demand in P-limited soils [35]. Co-inoculation of *Rhi. leguminosarum* and *Bacillus thuringiensis* was found to promote plant growth of *Pis. sativum* L. and *Lens culinaris* L (lentil). The enhancement in nodulation due to co-inoculations was 84.6% and 73.3% in pea and lentil, respectively, as compared to a single inoculation of *Rhi. leguminosarum*. The shoot dry weight gained on co-inoculations with variable cell populations of *Bac. thuringiensis* varied from 1.04 to 1.15 times and 1.03 to 1.06 times in pea and lentil, respectively, while root dry weight ratios of co-inoculations varied from 0.98 to 1.14 times and 1.08 to 1.33 times in pea and lentil, respectively, over a single inoculation of *Rhi. leguminosarum* [36]. Co-inoculation of *Rhizobium* sp. BHURC01 and *Pse. fluorescens* has been suggested to be an effective biofertilizer for *Cic. arietinum* production as it provided the maximum significant increases in nodule number, nodule dry weight, root dry weight, shoot dry weight, available N and available P [18]. Combined inoculation of *Rhizobium* sp*., Pse. fluorescens* and *Bac. megaterium* resulted in enhanced growth of *Vig. radiata* [37]. Co-inoculations with *Pseudomonas* spp. producing 1-aminocyclopropane-1-carboxylate (ACC) deaminase and *Rhi. phaseoli* enhanced plant growth and nodulation in *Vig. radiata* under salt-affected conditions [38]. Co-inoculation of *Ens. meliloti* (formerly *Sinorhizobium meliloti*) and *Exiguobacterium* sp. showed plant-growth promotion on *Trigonella foenum-graecum* (Fenugreek) with respect to the increases in root and shoot length, chlorophyll content, nodulation efficiency and nodule dry weight [39]. Co-inoculations of *Bradyrhizobium japonicum* USDA 110 with each of 10 strains of endophytic actinomycetes were determined for their potential to promote the growth of *Gly. max*. Co-inoculation of *Bradyrhizobium japonicum* USDA 110 with *Nocardia alba* conferred the maximum yield of root and shoot dry weight. Co-inoculations of *Bradyrhizobium japonicum* USDA 110 with each of *Noc. alba, Nonomuraea rubra* and *Actinomadura glauciflava* increased ARA up to 1.7-fold to 2.7-fold. For plant mineral composition, all of co-inoculation treatments significantly increased the nutrient levels of N, P, K, Fe, calcium (Ca), magnesium (Mg) and zinc (Zn) within soybean plants [40].

9.3. INFLUENCE OF INOCULANTS ON GENOTYPIC DIVERSITY OF INDIGENOUS RHIZOBIA

One of the important aspects of rhizobial populations in the field ecology is the characterization of existing soil strains that frequently compete with inoculant strains for nodule forming sites [41]. The effects of land management practices on rhizobial diversity have been suggested [42-44]. It has been suggested that agriculture creates highly selective and homogenous environments that reduce bacterial diversity [43]. Conversely, it has been argued that cultivation results in more diverse populations, and greater diversity of substrate utilization by total microbial communities from cultivated field than from pastures [42]. Palmer and Young [44] demonstrated that rhizobial genetic diversity could be higher in arable fields subjected to repeated cultivation than in relatively undisturbed grasslands. Other soil factors, such as pH, magnesium, heavy metals, clay, organic content and potential nitrogen, have been shown to influence rhizobial diversity [45-47]. The genetic diversity of rhizobia in relation to the host plant and the geographic origin has been studied. However, the results obtained from the previous studies have been argumentative. Laguerre *et al.* [48] evaluated the genetic diversity of 44 rhizobial isolates from different host plants and different geographic locations and reported that rhizobial classification at the genus and probably also the species level was independent of geographic origin and host plant affinity. Pongsilp *et al.* [49] reported that some of bradyrhizobial isolates from different soil samples in Thailand were closely related to each other. Lafay and Burdon [50] evaluated genetic structure of rhizobia nodulating *Acacia* spp. in southeastern Australia. The results showed that 2 genomospecies were both widespread and relatively abundant across the sampling sites, whereas another genomospecies seemed to be restricted to the more temperate regions. van Dillewijn *et al.* [51] determined the effect of *Ens. meliloti* on the rhizosphere microbial community of its host plant, *Medicago sativa* (alfalfa), and found that the plant appeared to have a much stronger influence on the microbial community as compared with *Ens. meliloti* inoculation. A number of studies suggest that the host plant and geographic origin are important factors affecting the genetic structure of rhizobial population. Bromfield *et al.* [52] reported the difference in the distribution of genotypes of *Ens. meliloti* from nodules of different legumes. Andronov *et al.* [53] found significant differences in the genetic structure between strains isolated from the soil under different legumes. Andronov *et al.* [54] found that isolates from a particular site belonged to a limited range of chromosomal genotypes. Carelli *et al.* [55] reported that the genetic structure of *Ens. meliloti* populations based mainly on differences among plants, while the effect of soil and plant cultivar were not significant. Pongsilp and Nuntagij [56] reported the genetic diversity among rhizobia was affected slightly by the host plants rather than the geographic origins.

9.4. APPLICATION OF RHIZOBIA IN BIOREMEDIATION

Rhi. leguminosarum strains and some other PGPR, such as *Pseudomonas aeruginosa* and *Serratia liquefaciens*, have been found to use crude oil, n-octadecane and phenanthrene as sole sources of carbon and energy. Co-inoculations of *Rhi. leguminosarum* with either *Pse. aeruginosa* or *Ser. liquefaciens* enhanced the phytoremediation potential of *Vicia faba* (faba bean) for oily desert sand through improving plant growth and N fixation [57]. *Rhi. leguminosarum* strains, *Bradyrhizobium japonicum* strains, *Pse. aeruginosa* and

Flavobacterium sp. could attenuate n-octadecane and phenanthrene in the surrounding nutrient medium. Intact nodules of *Vic. faba* containing these bacteria immobilized on and within those nodules reduced hydrocarbon levels in a medium [58].

REFERENCES

[1]   Zahran HH. *Rhizobium*-legume symbiosis and nitrogen fixation under severe conditions in an arid climate. Microbiol Mol Biol Rev 1999; 63: 968-89.

[2]   Yedegari M, Rahmani HA, Noormohammadi G, Ayneband A. Evaluation of bean (*Phaseolus vulgaris*) seeds inoculation with *Rhizobium phaseoli* and plant growth promoting rhizobacteria on yield and yield contents. Pak J Biol Sci 2008; 11: 1935-39.

[3]   Mathis JN, McMillin DE, Champion RA, Hunt PG. Genetic variation in two cultures of *Bradyrhizobium japonicum* 110 differing in their ability to impart drought tolerance to soybean. Curr Microbiol 1997; 35: 363-66.

[4]   Gao WM, Yang SS. A *Rhizobium* strain that nodulates and fixes nitrogen in association with alfalfa and soybean plants. Microbiol 1995; 141: 1957-62.

[5]   Pongsilp N, Nuntagij A. Selection and characterization of mungbean root nodule bacteria based on their growth and symbiotic ability in alkaline conditions. Suranaree J Sci Technol 2007; 14: 277-86.

[6]   Togay N, Togay Y, Cimrin KM, Turan M. Effects of *Rhizobium* inoculation, sulfur and phosphorus applications on yield, yield components and nutrient uptakes in chickpea (*Cicer arietinum* L.). Afr J Biotechnol 2008; 7: 776-82.

[7]   Erman M, Yildirim B, Togay N, Cig F. Effect of phosphorus application and *Rhizobium* inoculation on the yield, nodulation and nutrient uptake in field pea (*Pisum sativum* sp. *arvense* L.). J Anim Vet Adv 2009; 8: 301-04.

[8]   Ogutcu H, Kasimoglu C, Elkoca E. Effects of *Rhizobium* strains isolated from wild chickpeas on the growth and symbiotic performance of chickpeas (*Cicer arietinum* L.) under salt stress. Turk J Agric For 2010; 34: 361-71.

[9]   Zahir ZA, Yasin HM, Naveed M, *et al.* L-tryptophan application enhances the effectiveness of *Rhizobium* inoculation for improving growth and yield of mungbean [*Vigna radiata* (L.) Wilczek]. Pak J.Bot 2010; 42: 1771-80.

[10]   Kala TC, Christi RM, Bai NR. Effect of *Rhizobium* inoculation on the growth and yield of horsegram (*Dolichos biflorus* Linn). Plant Arch 2011; 11: 97-99.

[11]   Engelhard M, Hurek T, Reinhold-Hurek B. Preferential occurrence of diazotrophic endophytes, *Azoarcus* spp., in wild rice species and land races of *Oryza sativa* in comparison with modern races. Envir Microbiol 2000; 2: 131-41.

[12]   Lupwayi NZ, Clayton GW, Hanson KG, *et al.* Endophytic rhizobia in barley, wheat and canola roots. Can J Plant Sci 2004; 84: 37-45.

[13]   Sabry SRS, Saleh SA, Batchelor CA, *et al.* Endophytic establishment of *Azorhizobium caulinodans* in wheat. Proc Royal Soc London B 1997; 264: 341-46.

[14]   Singh RK, Mishra RPN, Jaiswal HK. Isolation and identification of natural endophytic rhizobia from rice (*Oryza sativa* L.) through rDNA PCR-RFLP and sequence analysis. Curr Microbiol 2006; 52: 117-22.

[15] Yanni YG, Rizk RY, Corich V, *et al.* Natural endophytic association between *Rhizobium leguminosarum* bv. *trifolii* and rice roots and assessment of its potential to promote rice growth. Plant Soil 1997; 194: 99-14.

[16] Mia, MAB, Shamsuddin ZH. *Rhizobium* as a crop enhancer and biofertilizer for increased cereal production. Afr J Microbiol 2010; 9: 6001-09.

[17] Antoun H, Beauchamp C, Goussard N, *et al.* Potential of *Rhizobium* and *Bradyrhizobium* species as plant growth promoting rhizobacteria on non-legumes: effect on radishes (*Raphanus sativus* L.). Plant Soil 1998; 204: 57-67.

[18] Nimnoi P, Pongsilp N. Genetic diversity and plant-growth promoting ability of the indole-3-acetic acid (IAA) synthetic bacteria isolated from agricultural soil as well as rhizosphere, rhizoplane and root tissue of *Ficus religiosa* L., *Leucaena leucocephala* and *Piper sarmentosum* Roxb. Res J Agric Biol Sci 2009; 5: 29-41.

[19] Gutierrez-Zamora ML, Martinez-Romero E. Natural endophytic association between *Rhizobium etli* and maize (*Zea mays* L.). J Biotechnol 2001; 91: 117-26.

[20] Pantujit S, Pongsilp N. Phosphatase activity and effects of phosphate-solubilizing bacteria on yield and uptake of phosphorus in corn. World Appl Sci J 2010; 8: 429-35.

[21] Chi F, Shen SH, Cheng HP, *et al.* Ascending migration of endophytic rhizobia, from roots to leaves, inside rice plants and assessment of benefits to rice growth physiology. Appl Envir Microbiol 2005; 7271-78.

[22] Biswas JC, Ladha JK, Dazzo FB. Rhizobia inoculation improves nutrient uptake and growth of lowland rice. Soil Sci Soc Am J 2000; 64: 1644-50.

[23] Husssain MB, Mehboob I, Zahir ZA, *et al.* Potential of *Rhizobium* spp. for improving growth and yield of rice (*Oryza sativa* L.). Soil Envir 2009; 28: 49-55.

[24] Engelhard M, Hurek T, Reinhold-Hurek B. Preferential occurrence of diazotrophic endophytes, *Azoarcus* spp., in wild rice species and land races of *Oryza satvia* in comparison with modern races. Envir Microbiol 2000; 2: 131-41.

[25] Chaintreuil C, Giraud E, Prin Y, *et al.* Photosynthetic bradyrhizobia are natural endophytes of the African wild rice *Oryza breviligulata*. Appl Envir Microbiol 2000; 66: 5437-47.

[26] Verma JP, Yadav J, Tiwari KN. Application of *Rhizobium* sp. BHURC01 and plant growth promoting rhizobacteria on nodulation, plant biomass and yields of chickpea (*Cicer arietinum* L.). Int J Agric Res 2010; 5: 148-56.

[27] Pradhan SK, Thatoi HN, Misra AK. Improvement of growth and N_2 fixation of *Stylosanthes hamata* (L.) taub. in wasteland soil through dual inoculation of *Rhizobium* and AM fungi with rock phosphate application. J Indian Bot Soc 2000; 79: 83-87.

[28] Rudresh DL, Shivapakash MK, Prasad RD. Effect of combined application of *Rhizobium*, phosphate solubilizing bacterium and *Trichoderma* spp. on growth, nutrient uptake and yield of chickpea (*Cicer aritenium* L.). Appl Soil Ecol 2005; 28: 139-46.

[29] Valverde A, Burgos A, Fiscella T, *et al.* Differential effects of coinoculations with *Pseudomonas jessenii* PS06 (a phosphate-solubilizing bacterium) and *Mesorhizobium ciceri* C-2/2 strains on the growth and seed yield of chickpea under greenhouse and field conditions. Plant Soil 2006; 287: 43-50.

[30] Anandham R, Sridar R, Nalayini P, *et al.* Potential for plant growth promotion in groundnut (*Arachis hypogaea* L.) cv. ALR-2 by co-inoculation of sulfur-oxidizing bacteria and *Rhizobium*. Microbiol Res 2007; 162: 139-53.

[31]   Rajendran G, Sing F, Desai AJ, Archana G. Enhanced growth and nodulation of pigeon pea by co-inoculation of *Bacillus* strains with *Rhizobium* spp. Bioresource Technol 2008; 99: 4544-50.

[32]   Elkoca E, Kantar F, Sahin F. Influence of nitrogen fixing and phosphate solubilizing bacteria on nodulation, plant growth and yield of chickpea. J Plant Nutr 2008; 33: 157-71.

[33]   Remans R, Croonenborghs A, Gutierrez RT, *et al.* Effects of plant growth promoting rhizobacteria on nodulation of *Phaseolus vulgaris* L. are dependent on plant P nutrition. Eur J Plant Pathol 2007; 119: 341-51.

[34]   Remans R, Ramaekers L, Schelkens S, *et al.* Effect of *Rhizobium–Azospirillum* coinoculation on nitrogen fixation and yield of two contrasting *Phaseolus vulgaris* L. genotypes cultivated across different environments in Cuba. Plant Soil 2008; 312: 25-37.

[35]   Ferrari AE, Wall JG. Coinoculation of black locust with *Rhizobium* and *Glomus* on a desurfaced soil. Soil Sci 2008; 173: 195-02.

[36]   Mishra PK, Mishra S, Selvakumar G, *et al.* Coinoculation of *Bacillus thuringeinsis*-KR1 with *Rhizobium leguminosarum* enhances plant growth and nodulation of pea (*Pisum sativum* L.) and lentil (*Lens culinaris* L.). World J Microbiol Biotechnol 2009; 25: 753-61.

[37]   Anandaraj B, Delapierre ALR. Studies on influence of bioinoculants (*Pseudomonas fluorescens, Rhizobium* sp., *Bacillus megaterium*) in green gram. J Biosci Tech 2010; 1: 95-99.

[38]   Ahmad M, Zahir ZA, Asghar HN, Asghar M. Inducing salt tolerance in mungbean through coinoculation with rhizobia and plant-growth-promoting rhizobacteria containing 1-aminocyclopropane-1-carboxylate deaminase. Can J Microbiol 2011; 57: 578-89.

[39]   Rajendran G, Patel MH, Joshi SJ. Isolation and characterization of nodule-associated *Exiguobacterium* sp. from the root nodules of fenugreek (*Trigonella foenum-graecum*) and their possible role in plant growth promotion. Int J Microbiol 2012; doi:10.1155/2012/693982

[40]   Nimnoi P, Pongsilp N, Lumyong S. Co-inoculation of soybean (*Glycine max*) with *Actinomycetes* and *Bradyrhizobium japonicum* enhances plant growth, nitrogenase activity and plant nutrition. J Plant Nutr (In press).

[41]   Date RA, Hurse LS. Intrinsic antibiotic resistance and serological characterization of populations of indigenous *Bradyrhizobium* isolated from nodules of *Desmodium intortum* and *Macroptilium atropurpureum* in three soils of SE Queensland. Soil Biol Biochem 1991; 23:551-61.

[42]   Kennedy AC, Smith KL. Soil microbial diversity and the sustainability of agricultural soils. Plant Soil 1995; 170:75-86.

[43]   Martinez-Romero E, Caballero-Mellado J. *Rhizobium* phylogenies and bacterial genetic diversity. Crit Rev Plant Sci 1996; 15:113-40.

[44]   Palmer KM, Young JPW. Higher diversity of *Rhizobium leguminosarum* biovar viciae populations in arable soils than in grass soils. Appl Envir Microbiol 2000; 66: 2445-50.

[45]   Harrison SP, Jones DG, Young JPW. *Rhizobium* population genetics: genetic variation within and between populations from diverse locations. J Gen Microbiol 1989; 135: 1061-69.

[46]   Labes G, Ulrich A, Lentzsch P. Influence of bovine slurry deposition on the structure of nodulating *Rhizobium leguminosarum* bv. viciae soil populations in a natural habitat. Appl Envir Microbiol 1996; 62:1717-22.

[47]   Hirsch PR, Jones MJ, McGrath SP, Giller KE. Heavy metals from past applications of sewage sludge decrease the genetic diversity of *Rhizobium leguminosarum* biovar trifolii populations. Soil Biol Biochem 1993; 25:1485-90.

[48]   Laguerre G, P. van Berkum P, Amarger N, Prevost D. Genetic diversity of rhizobial symbionts isolated from legume species within the genera *Astragalus*, *Oxytropis* and *Onobrychis*. Appl Envir Microbiol 1997; 63: 4748-58.

[49]   Pongsilp N, Teaumroong N, Nuntagij A, *et al.* Genetic structure of indigenous non-nodulating and nodulating populations of *Bradyrhizobium* in soils from Thailand. Symbiosis 2002; 33: 39-58.

[50]   Lafay B, Burdon JJ. Small-subunit rRNA genotyping of rhizobia nodulating Australian *Acacia* spp. Appl Envir Microbiol 2001; 67: 396-02.

[51]   van Dillewijn P, Villadas PJ, Toro N. Effect of a *Sinorhizobium meliloti* strain with a modified *putA* gene on the rhizosphere microbial community of alfalfa. Appl Envir Microbiol 2002; 68:4201-08.

[52]   Bromfield ESP, Behara AMP, Singh RS, Barran LR. Genetic variation in local populations of *Sinorhizobium meliloti*. Soil Biol Biochem 1998; 30: 1707-16.

[53]   Andronov EE, Roumyantseva ML, Simarov BV. Genetic diversity of a natural population of *Sinorhizobium meliloti* revealed in analysis of cryptic plasmids and IS*Rm2011-2* fingerprints. Russ J Genet 2001; 37: 494-99.

[54]   Andronov EE, Terefework Z, Roumiantseva ML, *et al.* Symbiotic and genetic diversity of *Rhizobium galegae* isolates collected from the *Galega orientalis* gene center in the Caucasus. Appl Envir Microbiol 2003: 69: 1067-74.

[55]   Carelli M, Gnocchi S, Fancelli S, *et al.* Genetic diversity and dynamics of *Sinorhizobium meliloti* populations nodulating different alfalfa cultivars in Italian soils. Appl Envir Microbiol 2000; 66: 4785-89.

[56]   Pongsilp N, Nuntagij A. Genetic diversity and metabolites production of root-nodule bacteria isolated from medicinal legumes *Indigofera tinctoria*, *Pueraria mirifica* and *Derris elliptica* Benth. grown in different geographic origins across Thailand. Amer-Eur J Agric Envir Sci 2009; 6: 26-34.

[57]   Radwan SS, Dashti N, El-Nemr IM. Enhancing the growth of *Vicia faba* plants by microbial inoculation to improve their phytoremediation potential for oily desert areas. Int J Phytoremediat 2005; 7: 19-32.

[58]   Radwan SS, Dashti N, El-Nemr IM, Khanafer M. Hydrocarbon utilization by nodule bacteria and plant growth-promoting rhizobacteria. Int J Phytoremediat 2007; 9: 475-86.

INDEX

A

1-aminocyclopropane-1-carboxylic acid (ACC) deaminase 53, 73, 77, 85, 176, 182

acetylene reduction activity (ARA) 75, 176-178, 181-182

agglutination (AG) 63

Agrobacterium 50-51, 97, 129, 142, 156, 159-160

Allorhizobium 5, 15, 133, 156

ammonification 82-83

amplified 16S rDNA restriction analysis (ARDRA) 119, 125-126, 131

amplified fragment length polymorphism (AFLP) 9, 107, 118-119, 131, 142, 144

angiosperm 4

anthranilate 82

antibody 49, 63

antigen 49, 63-64

antisera, antiserum 63-64

arbitary primer 108, 110

auxin 73, 78

Azorhizobium 3, 5, 6, 34, 55, 74, 81, 94, 116, 132, 157, 159-160, 166-167, 178-179

B

bacteroid 3, 53, 80, 82

biofertilizer 176-177, 180-181

biological nitrogen fixation (BNF) 3-4, 176

bioremediation 176, 183

Blastobacter 5, 26, 156

BOX primer 116-117

BOX sequence 113

BOXA1R 114, 115

BOX-PCR 107, 113-117, 119

bradyrhizobia, bradyrhizobial isolate, bradyrhizobial strain, *Bradyrhizobium* 3-8, 10, 26-29, 50-66, 74-76, 78-80, 82, 84, 94-95, 98, 109, 111, 114, 116-119, 132, 137, 141, 143-144, 156-157, 159-161, 163-167, 177-179, 182-183

Bradyrhizobiaceae 26, 156

Brucellaceae 35, 156
Burkholderia 3, 7, 35-36, 74, 85, 115, 156-157, 159-160
Burkholderiaceae 35

C

carboxylate 81-82
catecholate 81-82
cluster 52, 114-118, 127-129, 137, 156-157, 163
co-inoculation 176, 179-182
cryptic plasmid, cryptic megaplasmid 73, 77, 93-97
Cupriavidus 3, 7, 35, 54, 74, 84, 115, 156-157, 159-160, 166-167
cytokinin 78

D

dendrogram 51-52, 109-110, 116-117, 125-126, 128, 141, 143, 154, 157, 160
Devosia 3, 6, 34, 55-56, 60, 62, 156-157, 159-160

E

endonuclease 96, 118, 125-127, 129, 131, 136, 142
endophyte 26, 176, 178, 179
Ensifer 3, 5-7, 9-10, 16-17, 24-25, 51, 54-66, 73-75, 77, 79-82, 85, 93-94, 96-99, 109, 111-112, 116-117, 126, 132, 137-138, 140-144, 157, 159-160, 162-168, 177, 179, 182-183
enterobacterial repetitive intergenic consensus (ERIC) 107, 109, 113-117, 119, 127, 142, 144
enzyme pattern 9, 49, 52
enzyme-linked immunosorbent assay (ELISA) 63
ethylene 53, 75-76, 85-86

F

Fabaceae 4
fingerprint, fingerprinting 9, 95, 107, 109, 112-119, 125, 136-137, 139-144
fix gene 93, 138, 143, 164, 166, 170
fluorescent-antibody (FA) test 63

G

genetic diversity 9-10, 96, 107, 109, 113-116, 119, 125-126, 137-138, 143, 182-183
genotypic diversity 9, 93, 96, 107, 109, 113-114, 116, 119, 125, 136-137, 141, 154, 182
gibberellin (GA) 73, 80-81

H

Herbaspirillum 3, 7, 37, 74, 112, 156-157, 159-160
host specificity 3, 4, 8, 94
housekeeping gene 4, 143, 154, 162, 168
hybridization 4, 51, 116, 119, 126-127, 131, 136-142, 144
hydroxamate 82
Hyphomicrobiaceae 34

I

immunodiffusion (ID) 63
indole-3-acetic acid (IAA) 73, 78-80, 85, 166, 179
insertion sequence (IS) 136, 138-148, 166
intergenic spacer between 16S and 23S rRNA sequences (16S-23S IGS), 16S-23S rRNA intergenic spacer (IGS) 119, 125, 127-132, 154, 161

L

large subunit ribosomal RNA gene (23S rRNA gene, 23S rDNA) 125, 127-129, 131-132, 154, 159-161
leghaemoglobin 81
legume, leguminous plant, leguminous tree 3-11, 52, 73-75, 78-79, 81, 85, 95, 109, 115, 119, 128, 156, 166, 176-177, 179, 181, 183

M

megaplasmid 7, 77, 93-96, 98, 143, 166
Mesorhizobium 3, 5, 7-8, 30-33, 53-62, 74, 79, 84-86, 93-95, 97, 99, 109, 111-113, 117, 126, 129, 132-133, 139, 146, 156-157, 159-160, 162, 166, 169, 178-180
Methylobacteriaceae 34

Methylobacterium 3, 6, 34, 74, 156-157, 159-160

Microvirga 3, 6, 35, 156, 164-165

multilocus enzyme electrophoresis (MLEE) 126, 127, 131

multilocus sequence analysis (MLSA) 154, 161-163

multilocus sequence typing (MLST) 154, 162

N

neighbor-joining (NJ) method 126, 155, 157

nif gene 6-7, 93, 130, 132, 137-138, 141, 143, 163-164, 166, 168

nitrogen fixation (N fixation) 6-7, 52-53, 75-77, 80-82, 93-94, 102, 132, 137, 140, 143, 166, 168, 170, 176, 177, 179, 181, 183

nitrogenase 6, 52, 73-75, 81-82, 86, 137, 141, 163-164, 181

Nod factor 52, 78, 81, 137, 163

nod gene 6-7, 93, 130, 132, 137-139, 142-143, 146, 163-164, 166-168

nodulation 5-7, 9, 36, 53, 73-74, 77-78, 80-81, 85, 93-95, 97, 102, 132, 137-138, 143, 166-167, 169, 180-182

nol gene 138

non-legume 9, 21, 73, 79, 176, 178

numerical analysis 9, 49, 51

numerical taxonomy 4, 51, 126, 131

O

Ochrobactrum 3, 6, 35, 56-57, 112, 156-157, 159-160

Oxalobacteriaceae 37

P

phenotypic 4, 9, 49, 51-52, 76, 115, 154-155

phosphatase 54, 80, 84-85

phosphate solubilization 73, 84-85, 178

phosphate-solubilizing activity 73, 83-85

phosphate-solubilizing bacteria, phosphate-solubilizing bacterium (PSB), phosphate-solubilizing microorganisms 83-84

Phyllobacteriaceae 30, 156

Phyllobacterium 3, 6, 34, 142, 156-157, 159-160

phylogenetic 4-5, 7, 115-116, 127-129, 154-157, 161-163, 165

phylogeny, phylogenies 9, 126, 154-156, 161-163

phytohormone 73, 77-78, 85, 166, 176

phytoremediation 183

plant growth regulator 176

plant-growth-promoting (PGP) activities, plant-growth-promoting trait 73, 77, 176, 179

plant-growth-promoting rhizobacteria (PGPR) 73, 77, 176, 179-180, 183

plasmid profile 9, 73, 93, 95-96, 98

polymerase chain reaction (PCR) 9, 95-97, 107-110, 112-119, 125-131, 137, 142, 144, 155, 157, 161, 165-166

polymerase chain reaction-restriction fragment length polymorphism (PCR-RFLP) 9, 97, 114, 119, 125-131, 142

polymorphism 9, 96-97, 107, 109-110, 117-118, 125, 128, 131, 136, 142

primer 96, 107-118, 126-130, 155, 157-158, 161, 165

promiscuous 9, 74

Pseudomonadaceae 37

Pseudomonas 3, 7, 37, 81, 180-183

pseudonodule 5, 16, 25, 34, 37

R

Ralstonia 3, 7, 35, 74, 156-157, 159

random amplified polymorphic DNA (RAPD) 107-110, 112-113, 116-119, 138, 142, 144

rep gene 93, 96

repetitive extragenic palindromic (REP) 95, 107, 113-117, 119

repetitive sequence based PCR (rep-PCR) 107, 112-117, 119

restriction endonuclease digestion 96

restriction enzyme 118, 125, 136

restriction fragment length polymorphism (RFLP) 9, 96-97, 110, 114, 116, 118-119, 125-131, 136-139, 142-144

rhizobactin 82

Rhizobiaceae 4, 11, 156

Rhizobium 3-15, 17, 24-25, 27, 31-33, 50-64, 66, 73-77, 79-82, 84-85, 93-101, 109-117, 119, 126-130, 132-133, 137-144, 146-148, 156-157, 159-163, 166, 168-170, 177-181, 183

root exudate 8, 52, 95

S

sequence 4-5, 7, 9-10, 51, 93, 96, 107-110, 112-113, 115, 118-119, 126, 128, 130-131, 136-141, 144, 154-157, 159-162, 165-167

serocluster 49, 64

serogroup 49, 63-65

Shinella 3, 6, 26, 51, 112, 157, 159

siderophore 73, 78, 81, 82, 176, 179-180

Sinorhizobium 3-7, 10, 16-17, 24-25, 51, 54, 74, 79-80, 84-85, 93, 96, 98, 102, 111, 117, 126-127, 132, 140-141, 144, 156-157, 159, 162-163, 166, 168, 178, 182

small subunit ribosomal RNA gene (16S rRNA gene, 16S rDNA) 4-5, 7, 51, 108, 114, 119, 125-129, 131-132, 154-157, 159-163

symbioses, symbiosis 3-8, 53, 73-78, 81-82, 93-95, 114, 140, 143, 156, 166, 168-169, 177, 181

symbiosis island 7, 93-94, 169

symbiotic effectiveness 49, 73, 75-77, 176-177

symbiotic gene 3-7, 86, 93-94, 114, 116-117, 125, 130, 136-138, 143, 154, 163, 166

symbiotic performance 73-74, 76-77, 180

symbiotic plasmid (pSym) 7, 77, 93-102, 168-171

symbiotic promiscuity 3, 8

symbiotic variation 73, 76

T

two-primers random amplified polymorphic DNA (TP-RAPD) 107, 112

U

Unweighted pair group method with arithmetic mean (UPGMA) 52, 109, 118, 126, 128

urease 50, 62, 82-83

W

whole genome 116, 118-119, 165-167

www.ingramcontent.com/pod-product-compliance
Lightning Source LLC
Chambersburg PA
CBHW080020240326
41598CB00075B/478